Gas Turbine Diagnostics

Signal Processing and Fault Isolation

Gas Turbine Diagnostics

Signal Processing and Fault Isolation

RANJAN GANGULI

CRC Press
Taylor & Francis Group
Boca Raton   London   New York

CRC Press is an imprint of the
Taylor & Francis Group, an **informa** business

MATLAB® is a trademark of The MathWorks, Inc. and is used with permission. The MathWorks does not warrant the accuracy of the text or exercises in this book. This book's use or discussion of MATLAB® software or related products does not constitute endorsement or sponsorship by The MathWorks of a particular pedagogical approach or particular use of the MATLAB® software.

CRC Press
Taylor & Francis Group
6000 Broken Sound Parkway NW, Suite 300
Boca Raton, FL 33487-2742

First issued in paperback 2017

© 2013 by Taylor & Francis Group, LLC
CRC Press is an imprint of Taylor & Francis Group, an Informa business

No claim to original U.S. Government works

Version Date: 20121016

ISBN 13: 978-1-4665-0272-7 (hbk)
ISBN 13: 978-1-138-07442-2 (pbk)

Library of Congress Cataloging-in-Publication Data

Ganguli, Ranjan.
  Gas turbine diagnostics : signal processing and fault isolation / Ranjan Ganguli.
  pages cm
  Includes bibliographical references and index.
  ISBN 978-1-4665-0272-7 (hardback)
  1. Aircraft gas-turbines--Testing--Data processing. 2. Fault location (Engineering) 3. Airplanes--Monitoring. I. Title.

  TL709.G33 2013
  629.134'3530287--dc23   2012027103

Visit the Taylor & Francis Web site at
http://www.taylorandfrancis.com

and the CRC Press Web site at
http://www.crcpress.com

Contents

Preface

Gas turbines are very important components of modern infrastructure and are widely used in power generation. In particular, gas turbines are used for propulsion in jet engines that power most commercial and military aircraft. Faults in gas turbine engines can result in major problems, such as delays and cancellations of flights. Engine in-flight shutdowns (IFSDs) are particularly problematic and can have an impact on flight safety. Unscheduled engine removals add to the cost of air transport.

A systematic analysis of engine data has shown that most engine malfunction is preceded by a so-called single fault, which is a fault in one engine module or component. These single faults occur as sharp changes in measurement deviations in the jet engine, when compared to a baseline good engine. In this book, we present and illustrate a number of algorithms for fault diagnosis in gas turbine engines. These methods focus on the aspects of filtering or cleaning the measurement data and on fault isolation algorithms that use simple engine models for finding the type of fault in the engine. Novel methods for detecting the damage by finding the time location of a sudden change in the signal are also given. These methods include those based on Kalman filters, neural networks, and fuzzy logic and a hybrid soft computing approach.

The book provides a discussion of the different methods in data filtering, trend shift detection, and fault isolation developed over the past decade. Each method is demonstrated through numerical simulations that can be easily done by the reader using worksheets such as MS Excel or through MATLAB®. The book provides a variety of new research tools for use in the condition monitoring of jet engines. Though the measurements and models are specific to a turbofan engine, the algorithms given in this book will be useful to all engineers and scientists working on fault diagnosis of gas turbine engines. The data cleaning algorithms based on nonlinear signal processing shown in this book are also applicable to condition and health monitoring problems in general, and as in all such problems, sharp changes in measurement data herald the onset of a fault.

This book will be useful for engineers and scientists interested in gas turbine diagnostics. It will also be of interest to researchers in signal processing and those working on the fault isolation of systems. The algorithms presented in this book have broad appeal and can be used for condition and health monitoring of a variety of systems.

I acknowledge Dr. Allan Volponi and Hans Depold, Pratt & Whitney, who introduced me to the field of gas turbine diagnostics. I am grateful to my students Rajeev Verma, Niranjan Roy, Buddhidipta Dan, Payuna Uday,

V.N. Guruprakash, and V.P. Surendar for testing the algorithms and generating the numerical results. I am also grateful to K. Bhanu Priya for helping typeset the document. Finally, I am grateful to the Indian Institute of Science for furnishing an ambient atmosphere for doing research.

Prof. Ranjan Ganguli
Bangalore

MATLAB® is a registered trademark of The MathWorks, Inc. For product information, please contact:

The MathWorks, Inc.
3 Apple Hill Drive
Natick, MA 01760-2098 USA
Tel: 508-647-7000
Fax: 508-647-7001
E-mail: info@mathworks.com
Web: www.mathworks.com

About the Author

Dr. Ranjan Ganguli is a professor in the Aerospace Engineering Department of the Indian Institute of Science (IISc), Bangalore. He received his MS and PhD degrees from the Department of Aerospace Engineering at the University of Maryland, College Park, in 1991 and 1994, respectively, and his BTech degree in aerospace engineering from the Indian Institute of Technology in 1989. He worked in Pratt & Whitney on engine gas path diagnostics during 1998–2000. During his academic career at IISc since 2000, he has conducted sponsored research projects for companies such as Boeing, Pratt & Whitney, Honeywell, HAL, and others. He has published over 140 papers in refereed journals and has presented over 80 papers in conferences. He has published books entitled *Structural Health Monitoring Using Genetic Fuzzy Systems* and *Engineering Optimization*. He is a fellow of the American Society of Mechanical Engineers, a fellow of the Royal Aeronautical Society, an associate fellow of the American Institute of Aeronautics and Astronautics, and a fellow of the Indian National Academy of Engineering. He also received the Alexander von Humboldt Fellowship and the Fulbright Fellowship in 2007 and 2011, respectively. He is an associate editor of the *AIAA Journal* and of the *Journal of the American Helicopter Society*.

1

Introduction

Diagnostics of gas turbine engines is important because of the high cost of engine failure and the possible loss of human life. In this book, we will focus on aircraft or jet engines, which are a special class of gas turbine engines. Typically, physical faults in a gas turbine engine include problems such as erosion, corrosion, fouling, built-up dirt, foreign object damage (FOD), worn seals, burned or bowed blades, etc. These physical faults can occur individually or in combination and cause changes in performance characteristics of the compressors, and in their expansion and compression efficiencies. In addition, the faults cause changes in the turbine and exhaust system nozzle areas. These changes in the performance of the gas turbine components result in changes in the measurement parameters, which are therefore dependent variables. This chapter introduces some basic concepts that are necessary for an understanding of gas turbine diagnostics. First, the importance of signal processing in noise removal from measurements is highlighted. Next, the typical gas turbine diagnostic process is explained. The widely used linear filters and the median filter are then introduced. This is followed by an outline of the least-squares approach and the Kalman filter. Finally, the role of influence coefficients and the basics of vibration-based diagnostics are highlighted.

1.1 Background

Many problems in jet engines manifest themselves as changes in the gas path measurements [1–3]. Typical gas path measurements are exhaust gas temperature (*EGT*), low rotor speed (*N1*), high rotor speed (*N2*), and fuel flow (*WF*). These measurements are also called cockpit parameters, as they are displayed to the pilot. Some newer engines also have additional pressure and temperature probes between the compressors and turbines. However, the cockpit parameters are present in both newer and older engines, and therefore fault detection and isolation systems should be able to work for older engines, which are more susceptible to damage. Jet engine gas path analysis works on deviations in gas path measurements from an undamaged baseline engine to detect and isolate faults. These deviations in the measurements from baseline are known as measurement deltas and are plotted vs. time, and the resulting computer graphics (known as trend plots) are used by power plant engineers to visually analyze the condition of the engine and its different modules. Unfortunately, noise contaminates the measurement deltas, thereby reducing the signal-to-noise ratio. This can hide key features in the signal from a person observing the data. A key objective of gas turbine diagnostics is to make decisions about the existence and location of faults from the noisy data.

A typical measurement delta has two main features. The first is because of long-term deterioration that can be considered to vary in time as a low-degree polynomial, with a linear approximation being very satisfactory [4, 5]. The second feature of the measurement delta is sudden step-like changes due to so-called single faults. Depold and Gass [6] conducted a statistical study of airline data and discovered that the main cause of many engine in-flight shutdowns was these single faults, which were preceded by a sharp change in one or more of the measurement deltas. Such a sharp trend change can also happen if the engine is repaired and tested on the ground in a test cell. Therefore, a typical jet engine measurement delta signal can be assumed to be a linear long-term deterioration along with sudden step changes due to a single-fault or a repair event.

The power plant engineer does not solely rely on observing trend plots to monitor the engine condition. Various diagnostic algorithms have been developed to estimate engine condition and identify faults from the health signals using weighted least squares [7, 8], Kalman filter [9], neural network [6, 10–12], fuzzy logic [13], and Bayesian [14] approaches. However, while all these algorithms attempt to handle uncertainty in the measurement deltas, their performance is often degraded as the noise in the data increases. This is also true for system identification of jet engines [15] that is done to produce better control and diagnostics models. In addition, these estimation and pattern recognition algorithms are often optimal for Gaussian noise models and can degrade when non-Gaussian outliers are present in the data [16].

Classical signal processing has been dominated by the assumption of a Gaussian random noise model for defining the statistical properties of a real process. However, many real-world processes are characterized by impulsive noise that causes sharp spikes and outliers in the data. For example, data can be corrupted by impulsive noise during acquisition and transmission through communication channels [17]. Phenomena such as atmospheric noise is also impulsive in nature. Fault detection and isolation methods that are optimized for random Gaussian noise can suffer severe performance degradation under non-Gaussian noise. Therefore, signal processing of the measured data can be very useful for improving gas turbine diagnostics. In particular, impulsive noise should be removed.

1.2 Signal Processing

In signal processing, filtering methods are used to preprocess the data to reduce noise. The term *noise* here is used in a general sense and includes any corruption to the signal that hinders the pattern recognition or state estimation process or leads to false artifacts being observed during visualization. Traditionally, smoothing methods used by the gas turbine industry are moving averages and exponential smoothing [6]. The moving average is a special case of the finite impulse response (FIR) filter, and the exponential average is a special case of the infinite impulse response (IIR) filter. These filters will be explained later in this chapter. Depold and Gass [6] first addressed the problem of finding a filter that preserves the sharp trend shifts in gas path measurements due to a single fault. They showed that the exponential average filter has a faster reaction time than the widely used 10-point average and is therefore a better filtering method for processing data prior to trend detection and fault isolation. They also developed some rules of thumb to remove outliers from gas turbine measurements. These rules were based on the logic that a shift in any one measurement without shifts in the other measurements would indicate an outlier.

However, both the FIR and IIR filters are linear filters and remove noise while blurring the edges in the signal. In addition, the human visual system is acutely sensitive to high frequency in the spatial form of edges [18]. Most of the low frequency in an image is discarded by the visual system before it can even leave the retina. Unfortunately, the presence of sporadic high-amplitude impulsive noise in a signal can confuse the human visual system into seeing patterns where none are really present. Such noise can also trigger an automated trend detection system to give a false alarm. Therefore, it is necessary to remove any such high-amplitude noise while preserving edges from the measurement deltas before subsequent data processing operations for fault detection and isolation.

Substantial research efforts have been conducted in the field of image processing to find suitable alternatives to linear filters that are robust or resistant to the presence of impulsive noise. Among these works, the approach that has received the most attention is that of median filters. Median filters are a well-known and useful class of nonlinear filters in the image processing field [19–24]. They are useful for removing noise while preserving fine details in the signal. However, they are not well known in engineering health monitoring applications. Ganguli [25] used FIR-median hybrid (FMH) filters [20] for removing noise from gas turbine measurements while preserving trend shifts. In this study, step changes were considered in a constant signal as a representation of a single-fault event. Results showed that the FMH filter preserved the sharp trend shifts in the signal while the moving average and exponential average filter smoothed the trend shifts. The problem of deterioration was not addressed. Furthermore, the FMH filter used in this study required up to 10 points of forward data and therefore had a 10-point time lag. Since jet engines often get only 1 or 2 points in each flight, the 10-point time lag is very large and is more suitable for engines with online diagnostics systems or for systems where data are obtained rapidly. The cost of high-rate data acquisition remains quite high. In applications other than gas turbine engines, Nounou and Bakshi [26] used the FIR-median hybrid (FMH) filter to remove noise from chemical process signals. Manders et al. [27] used a median filter of length 5 to remove noise in temperature data for monitoring the cooling system of an automobile engine having installed thermocouples and pressure sensors. Ogaji et al. [28] used FMH filters to remove noise from data measured by a global positioning system (GPS) that directly measures relative displacement and position coordinates for a tall building.

Nonlinear filters are not limited to median type filters. A special class of neural networks called the autoassociative neural network (AANN) [29, 30] has been used for noise filtering, using sensor replacement and gross error detection and identification. Lu et al. [11, 31] used autoassociative neural networks for noise filtering gas path measurements. The AANN performs a unitary mapping, which maps the input parameters onto themselves. The AANN is also capable of removing any outliers in the data, and performed better at preserving trend shifts than the moving average or exponential average filter. To train the AANN, noisy data are input to it and mapped to noise-free data at the output nodes. The number of input nodes and output nodes is equal to the number of measurements. The AANN has an input and output layer, two hidden layers, and a bottleneck layer. Thus, the data go to the input layer, then a hidden layer, then a bottleneck layer, followed by a hidden layer and the output layer. Lu et al. [11] used eight measurement nodes for the hidden layer and five nodes for the bottleneck layer, resulting in an 8-9-5-9-8 AANN architecture. The neural network therefore learns the noise characteristics of the data and is trained to give noise-free data from noisy data. We will discuss the AANN in more detail in Chapter 9.

Many filtering algorithms use a fixed-noise detection threshold obtained at a presumed noise density level. For example, wavelet-based noise removal methods [26, 32, 33] use orthogonal wavelet analysis, which finds coefficients related to undesired features in the signal. Nounou and Bakshi [26] showed that wavelet-based noise removal methods could be superior to the FMH filter for processing signals with sharp trend shifts. The wavelet-based noise removal has three parts: (1) orthogonal wavelet transform, (2) thresholding of wavelet coefficients, and (3) inverse wavelet transform. By setting to zero the wavelet coefficients at the highest orthogonal level of decomposition, noise can be removed from the signal. However, finding a threshold depends on the noise level and nature of the noise and is a difficult problem. Neural network-based filtering methods are also sensitive to the noise levels in the training data. For example, the AANN used by Lu et al. [11] was trained with representative noisy data using simulated signals. However, when the noise characteristic becomes different from that used in algorithm development, which can happen in practical applications, the performance of these algorithms can show degradation.

1.3 Typical Gas Turbine Diagnostics

Urban [34] states the scope of gas turbine diagnostics in his research paper as follows: "Therefore, it follows that if physical problems result in degraded component performance, which in turn produce changes in the measurable engine parameter, then it is possible to utilize these measurable changes to isolate the degraded component characteristics, in whatever combination, and permit correction of the causative problems."

Figure 1.1 shows a schematic representation of the gas turbine diagnostics process. The measurement deltas are processed using smoothing algorithms based on moving or exponential averages [6]. In some cases, the diagnostics function may be completely performed by power plant engineers. In these cases, the measurement deltas are visualized using computer graphics and the power plant engineer uses his or her experience to detect engine deterioration or faults. In case a fault or severe performance degradation is detected, the power plant engineer may suggest prognostics and maintenance action. In other cases, the power plant engineer may also have access to automated fault detection and isolation software that can estimate the condition of the different modules and also detect and isolate other faults. In addition, expert systems may be available for interpreting

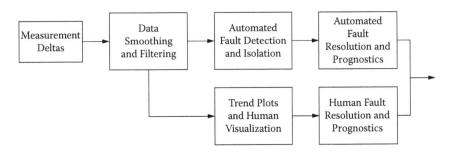

FIGURE 1.1
Schematic representation of gas turbine diagnostics process. (From Ganguli, R., *Journal of Propulsion and Power* 19(5):930–937, 2003. With permission.)

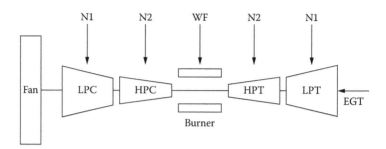

FIGURE 1.2
Schematic representation of gas turbine engine modules and sensor measurements. (From Ganguli, R., *Journal of Propulsion and Power* 19(5):930–937, 2003. With permission.)

the output of the fault detection and isolation algorithms for suggesting maintenance and prognostics action. In general, both the automated and human components of the diagnostics system should be used for the best possible decisions.

Figure 1.2 shows a schematic of a turbo engine that has five modules: fan, low-pressure compressor (LPC), high-pressure compressor (HPC), low-pressure turbine (LPT), and high-pressure turbine (HPT). Air is sucked into the engine through the fan and compressed in the LPC and HPC. Then, the compressed air is mixed with a fuel and burned in the burner. Following this, the hot gases are passed through the turbines and power is generated during this process. Finally, the hot gases are sent out through the exhaust.

Faults in the gas turbine engine cause efficiency deterioration for the engine modules. The engine state is monitored using at least the four basic sensors: exhaust gas temperature (*EGT*), fuel flow (*WF*), low rotor speed (*N*1), and high rotor speed (*N*2). The measurements that are taken at altitude at a given temperature are then converted to standard day sea level conditions, and then the baseline measurement of an undamaged engine at the same condition (usually from a thermodynamics-based performance model) is subtracted from the measurements to yield the measurement deltas Δ*EGT*, Δ*WF*, Δ*N*1, and Δ*N*2. The measurement deltas are then used for estimating the engine state. Various fault isolation algorithms are used to find the module where the fault has occurred. These include Kalman filter, neural networks, and fuzzy logic-based methods, some of which will be discussed in later chapters.

We can observe from Figure 1.1 that a key component of the diagnostics system is the smoothing or filtering function. While much research has been expended on the fault detection and isolation function, not much work has been done to improve the data smoothing and filtering function [6, 11, 25, 31]. The next two sections give a brief background on linear filters and the non-linear median filter. Several variations of the median filter will be discussed in this book for application to gas turbine diagnostics.

1.4 Linear Filters

The finite impulse response (FIR) filter can be represented as

$$y(k) = \sum_{i=1}^{N} b(i)x(k-i+1) \tag{1.1}$$

where $x(k)$ is the kth input measurement and $y(k)$ is the kth output. N is the filter length and $\{b(i)\}$ is the sequence of weighting coefficients, which define the characteristics of the filter and sum to unity. When all the weights $\{b(i)\}$ are equal, the FIR filter reduces to the special case of the mean or average filter, which is widely used for data smoothing. For example, the 10-point moving average has the form

$$y(k) = \frac{1}{10}(x(k) + x(k-1) + x(k-2) + \cdots + x(k-9)) \tag{1.2}$$

Each of the 10 weights for this filter is equal to 1/10.

Exponentially Weighted Moving Average (EWMA) is a popular IIR filter that smoothes a measured data point $x(k)$ by exponentially averaging it with all previous measurements $y(k-1)$.

$$y(k) = ax(k) + (1-a)y(k-1) \tag{1.3}$$

The parameter a is an adjustable smoothing parameter between 0 and 1 with values such as 0.15 and 0.25 being routinely used in applications [6]. The exponential average filter has memory since it retains the entire time history by using the output of the last point. While linear filters are often used to smooth data before fault diagnosis, they can also smooth out important signal features. This problem is alleviated by the use of nonlinear filters such as the median filter.

1.5 Median Filters

Several median type filters are discussed in this book in Chapters 2–4, 6, and 7. Here, we introduce the standard median filter, which is well known in image processing.

Standard median (SM) filters are a popular and useful class of nonlinear filters. The success of median filters is based on two properties: edge preservation and noise reduction with robustness against impulsive type noise. Neither property can be achieved by traditional linear filtering without using

time-consuming and often *ad hoc* data manipulation. The median filter having length or window of $N = 2K + 1$ can be represented as [19]

$$y(k) = median(x(k - K), x(k - K + 1), ..., x(k), ..., x(k + K - 1), x(k + K)) \quad (1.4)$$

where $x(k)$ and $y(k)$ are the kth sample of the input and output sequences,
respectively. To compute the output of a median filter, an odd number of
sample values are sorted and the median value is used as the filter output.
The median filter thus uses both past and future values of $x(k)$ for predicting
the current output point. The above filter for discrete time k and window
length $N = 2K + 1$ can be written in compact form as

$$y = median(x_{-k}, ..., x_{-1}, x_0, x_1, ...x_k) \quad (1.5)$$

Since the output of a median filter is always one of the input samples, it is
possible that certain signals can pass through the median filter without being
altered. This has been shown to hold for median and many median-based filters. Since such signals define the nature of a filter, these are referred to as a root
signal. A root is a signal that is not modified by further filtering. Thus, a signal
is a root signal of the SM filter in Equation (1.5) if for all signal values it satisfies

$$x_0 = median(x_{-k}, ..., x_{-1}, x_0, x_1, ...x_k) \quad (1.6)$$

Repeated median filtering of any finite length signal will result in a root signal after a finite number of passes. It has been shown that if an SM filter has filter
window width $2K + 1$ and the signal has length P, then at most $3[(P - 2)/2(K + 2)]$
passes of the filter are required to produce a root signal [35]. However, this
bound is rather conservative in practice. Typically, after 5–10 filtering passes
only slight, if any, changes take place in the filter output and the filter is said to
have converged. Some of the filters discussed in the following chapters address
this convergence problem of the median and accelerate the signal processing.
    It is important to determine if a filter will drive any input signal to one of
these roots after a sufficient but finite number of passes. If it does, the filter is
said to have convergence property. The important fact is that the step edges,
ramp edges of sufficient extent, and constant regions are root signals of the
median filter. This means that such signals are preserved even after repeated
filtering, which is very important to the feature preservation property of the
median type filters. Note that step edges are typical of faults in gas turbine
measurements, and ramp edges are typical of long-term deterioration trends
in engines. Therefore, median type filters are well suited to preprocess gas turbine measurements. We have used the word *preprocess* to highlight the function of signal processing algorithms in gas turbine diagnostics. The processing
of the measurements to extract information about the engine state is typically
performed by least-squares and Kalman filter type algorithms. Software
packages based on these algorithms have been developed by gas turbine
manufacturers. These algorithms are discussed in the next two sections.

1.6 Least-Squares Approach

The mathematics of gas path analysis ranges from being relatively simple to very sophisticated. Probably the simplest approach involves the weighted least-squares approach propounded by Doel [7, 36]. This approach is used in General Electric's TEMPER software and is discussed below.

The measurement process can be mathematically written as

$$z = h(x) + v \qquad (1.7)$$

where z is a measurement vector and x is a state vector. For example, fuel flow is a typical measurement and compressor efficiency is a typical state. The nonlinear relation between x and z is captured by $h(x)$. If the measurement is error-free, i.e., there is no error, then

$$z = h(x) \qquad (1.8)$$

The problem of finding x given z is then a typical inverse problem. In reality, the diagnostics problem is complicated by the presence of noise, and thus v is added as a vector of random error. The inverse problem then becomes more complicated and difficult to solve. Inverse problems with noise are similar to pattern recognition problems in many ways.

In gas turbine diagnostics, and in many other problems in engineering, a key simplification involves linearization. Thus, we can write

$$z = Hx + v \qquad (1.9)$$

Here H is a matrix and x is a vector. To make the mathematics simpler, we typically assume that the measurement error is Gaussian in nature. Also, the mean of the error is assumed to be zero, leading to zero mean white noise. The Gaussian assumption is also made for the state vector x. Since x and z are defined as deviations from baseline condition, this assumption is reasonable if suitable data are used.

We now define the covariance matrix of the state vector as

$$P = E(xx^T) \qquad (1.10)$$

and of the measurement error as

$$R = E(vv^T) \qquad (1.11)$$

where E is the expected value operator.

We also assume that the measurement error is statistically independent of the engine state:

$$E(xv^T) = 0 \qquad (1.12)$$

The optimal estimate of the state can now be found from the measurements by minimizing the quadratic form:

$$J = \frac{1}{2}\left\{ x^T P^{-1} x + (z - Hx)^T R^{-1}(z - Hx) \right\}$$

(1.13)

The optimal state vector is then obtained by setting $dJ = 0$ for an arbitrary dx^T:

$$\hat{x} = \left(P^{-1} + H^T R^{-1} H \right)^{-1} H^T R^{-1} z$$

(1.14)

The above approach is a weighted least-squares approach as the matrices P and R are used as weights to bring in the probabilistic nature of the system. These matrices are very important and some key statements need to be made about them.

1. The diagonal of R contains the variance of the measurement errors.
2. The off-diagonal elements of R contain the covariance between the measurements and are typically assumed to be zero.
3. Most off-diagonal elements of P are assured to be zero. However, some elements are likely to be nonzero. For instance, the fan flow capacity and fan efficiency are typically related.

The matrices H, R, and P are crucial for gas path analysis. These matrices need to be available for the gas path analysis to yield results once the measurement z is obtained. The weighted least-squares method has a tendency of smearing the effect of a measurement over several states. For example, consider a situation where there is 1% deterioration in the high-pressure turbine efficiency. This is of course an idealized and simulated situation where no other changes were present. The ideal measurements can be obtained using $z = Hx$. However, when the least-squares method is applied with the measurement z, the efficiency change will be distributed over other modules and components due to measurement uncertainty. Another problem with the least-squares approach is that engine modules or states that are not modeled will be assigned to modeled components. The sensitivity of the least-squares algorithm depends on the relative magnitude of the P and R matrices. This feature of dependence on the probability matrices is common of the gas path analysis algorithms. A good knowledge of the measurement statistics is needed for the algorithm to perform well. Also, since a linear model is assured between z and x, the algorithm is valid only when measurement deviations are small.

There are two main situations in which gas path analysis is used. They are on-wing on the airplane and in the test cell on the ground. In the on-wing situation, the data acquisition rate can range from a minimum of once per flight to much more regular intervals, such as every flight hour.

Some modern engines may have even faster rates of data acquisition. The advantage of on-wing monitoring is the ability to use the time history of the measurements. A test cell analysis, in contrast, is a snapshot analysis and does not yield this time history.

One way to include this time history is to use the Kalman filter, which we discuss in the next section. However, this feature can be incorporated in the least-squares approach by using smoothed analysis results for module deterioration and sensor error to give *a priori* estimates to analyze the new data. In TEMPER, exponential smoothing is used on the module deterioration and sensor error analysis results. Note that exponential smoothing is the IIR filter discussed earlier, and this filter has memory.

There is a key difference between the on-wing and the test cell. In the on-wing case, the measurement delta compares the present measurement value with the corresponding value in the recent past. On the other hand, for the test cell case, the measurement delta is the difference between the current measurement and a fixed baseline engine. Therefore, the on-wing case compares the engine to itself, while in the test cell case, the engine is compared to a set of similar engines.

There is one situation in gas path analysis that needs special mention. Sometimes, a sensor can show a large sudden change from its baseline value. Also, there can be a large shift in a module component. For example, foreign object damage can cause a large shift in a single component. Any large change in either the measurement or the module performance will violate the least-squares assumptions. Therefore, such cases will result in a large solution residual. To salvage this situation, a single cause of the large residual can be found. For example, TEMPER uses this approach if the solution residual becomes greater than the 95% confidence limit. This algorithmic approach is known as fault logic.

When fault logic is activated, a new weighted least-squares analysis is conducted for each sensor error and module fault. The standard deviation of the sensor or the module being considered is increased by 100%. If we have identified the correct fault, the solution residual will suddenly come back to normal range. This approach can alleviate one of the shortcomings of the least-squares analysis. The reader will observe that most of the complication in gas path analysis is caused by sensor error. However, since sensor error is realistic and inevitable, the gas turbine diagnostic algorithms must address this issue. A key risk associated with gas path algorithms lies in the possibility of misdiagnosis or false alarms. Inappropriate and unnecessary maintenance action can be triggered by such results. Doel [7] goes on to suggest that "the use of emerging technologies such as expert systems, fuzzy logic and neural networks might generate further gains." These will be discussed in later chapters. While the least-squares method is used in the TEMPER software of GE, the software that Pratt & Whitney created for engine health monitoring typically uses a type of Kalman filter. We introduce the Kalman filter in the next section.

1.7 Kalman Filter

The Kalman filter was developed in the 1960s and was used in gas turbine diagnostics in the late 1970s. Broadly speaking, there are two types of Kalman filters: discrete time and continuous time. The Kalman filter can be viewed as a generalization of the weighted least-squares approach discussed in the previous section.

The Kalman filter is an optimal estimator and estimates states of a linear dynamical system perturbed by Gaussian noise. Much of the development of the Kalman filter application for gas turbine diagnostics was done by Volponi and coworkers [9, 12]. We will discuss this work briefly in the current section. Consider the relationship between the measurement deviations and the state deviations:

$$z = Hx + v$$

We assume that x and v are independent and Gaussian. The optimal estimation problem is equivalent to minimizing

$$J = \frac{1}{2}\left\{(z - Hx)^T R^{-1}(z - Hx) + (x - \mu_x)^T P^{-1}(x - \mu_x)\right\} \tag{1.15}$$

We can see that the quadratic form here has a structure very similar to that given before in the section of least squares. The first term is a measure of measurement error and the second term is a measure of state error. The first term in Equation (1.15) is weighted by the inverse of the noise covariance, and the second term is weighted by the inverse of the state covariance. The optimal estimator is then given by

$$\hat{x} = \left[P_0^{-1} + H^T R^{-1} H\right]^{-1}\left(H^T R^{-1} z + P_0^{-1}\mu_x\right) \tag{1.16}$$

The optimal estimator can be put in predictor-corrector form:

$$\hat{x} = \mu_x + \left[P_0^{-1} + H^T R^{-1} H\right]^{-1} H^T R^{-1}\left[z - H\mu_x\right]$$
$$\hat{x} = \mu_x + P_0 H^T \left[H P_0 H^T + R\right]^{-1}\left[z - H\mu_x\right] \tag{1.17}$$

Here, the first term (μ_x) is the predictor and the second term is the corrector. The term $[z - H\mu_x]$ is called the residual. The term $P_0 H^T [H P_0 H^T + R]^{-1}$ is called the gain. Writing in a generalized form,

$$\hat{x} = \bar{x} + D(z - H\bar{x}) \tag{1.18}$$

We are now ready to formally define the discrete time Kalman filter. Consider a given discrete time k; then the system model is given as

$$x_k = \varphi(k)x_{k-1} + \omega_k \qquad (1.19)$$

where x_k is the state vector at discrete time (also called epoch) k, $\varphi(k)$ is the state transition matrix, and ω_k is the process noise vector.

Along with the system model, there exists the measurement model given by

$$z_k = H_k x_k + v_k \qquad (1.20)$$

Here z_k is the measurement vector, H_k is the geometry matrix, and v_k is the measurement noise vector. The following assumptions are made:

1. v_k and ω_k are Gaussian and zero mean.
2. $R_k = \text{cov } (v_k, v_k) > 0$.
3. $Q_k = \text{cov } (\omega_k, \omega_k) \geq 0$.
4. $\text{Cov } (\omega_k, v_j) = 0$; i.e., there is no correlation between process and measurement noise.
5. $\mu_x = E(x_0)$, or the initial guess of the state is known.
6. $P_0 = \text{cov } (x_0, x_0) = P_0 > 0$.

The discrete time Kalman filter equations are then given by five equations given below:

$$\hat{x}(k+1|k) = \varphi \ (k+1)\hat{x}_k$$

$$P(k+1|k) = \varphi \ (k+1) P_k \varphi^T (k+1) + Q_{k+1}$$

$$D_{k+1} = P(k+1|k) H_{k+1}^T \left(H_{k+1}^T P(k+1|k) H_{k+1}^T + R_{k+1} \right)^{-1} \qquad (1.21)$$

$$\hat{x}_{k+1} = \hat{x}(k+1|k) + D_{k+1} \left(z_{k+1} - H_{k+1} \ \hat{x}(k+1|k) \right)$$

$$P_{k+1} = [1 - D_{k+1} H_{k+1}] \ P(k+1|k)$$

These equations were stated in the above form by Volponi and form the basis of the application of the Kalman filter to gas path analysis. The first of the five equations represents an extrapolation of the state vector from the kth epoch to the $(k + 1)$th epoch. The transition matrix acts as an operator for this extrapolation. The second equation shows the extrapolation of the covariance matrix P from the kth epoch to the $(k + 1)$th epoch. The third equation involves calculation of the Kalman gain. The fourth equation represents the state update, and the fifth equation represents the covariance update.

The abstract notation of the Kalman filter will become clearer when we use it for the typical engine performance diagnostic application. This will be done in Chapter 8, where issues of multiple faults, single faults, and sensor faults will be discussed. The Kalman filter appears to be a precise mathematical construct. However, being a predictor-corrector method, it works best when the initial or guess estimate is close to the actual answer. In this behavior, it is similar to many numerical methods, such as the Newton-Raphson method. Also, the Kalman filter performance depends on the matrices P, Q, and R. These so-called numerics need to be selected judiciously for it to perform well.

However, the appropriate selection of these numerics is a nontrivial problem. Fortunately, the H matrix is available for gas turbine engines from the engine model. This matrix contains the very important and useful influence coefficients and is discussed in the next section.

1.8 Influence Coefficients

In gas turbine diagnostics, we typically want to know the state vector containing the engine fault delta. Typically, there are two parameters for a given module, which results in a total of 10 states. Thus, for the fan module, there is a change in fan efficiency and flow capacity. The compressor modules, i.e., the LPC and HPC, also have changes in efficiency and flow capacity associated with them. On the other hand, the turbine modules LPT and HPT have efficiency and an area associated with them. All 10 states are deltas or changes from a baseline position or value of the state. This engine state vector can be written as

$$x_e = \begin{Bmatrix} \Delta\eta_{FAN} \\ \Delta FC_{FAN} \\ \Delta\eta_{LPC} \\ \Delta FC_{LPC} \\ \Delta\eta_{HPC} \\ \Delta FC_{HPC} \\ \Delta\eta_{HPT} \\ A_4 \\ \Delta\eta_{LPT} \\ A_5 \end{Bmatrix}$$

Here η refers to the efficiency of the modules and FC to the flow capacity. Also, A_4 and A_5 are area changes associated with the HPT and the LPT. Similarly, there are measurement deltas. Generally, it is a good idea to work with percent deltas, as this avoids the use of physical variables that may have different units. The use of 1% measurement deltas automatically nondimensionalizes and normalizes the number.

$$z = \left\{ \begin{array}{c} \Delta N1 \\ \Delta N2 \\ \Delta WF \\ \Delta T3 \\ \Delta P3 \\ \Delta T25 \\ \Delta P25 \end{array} \right\}$$

Here $\Delta N1$ and $\Delta N2$ are the low and high spool speed deltas, ΔWF is the change in the fuel flow, $\Delta T3$ and $\Delta P3$ are the changes in the HPC exit temperature and pressure, and $\Delta T25$ and $\Delta P25$ are the deltas of temperature and pressure between the LPC and HPC. Each measurement delta is normalized as follows:

$$\Delta N1 = 100 \frac{N1_{mean} - N1_{base}}{N1_{base}} \tag{1.22}$$

These changes are from a baseline good engine. We now come to the engine fault influence coefficient matrix, which is defined in

$$z = H_e x_e \tag{1.23}$$

Here the subscript e refers to engine fault, compared to sensor fault, which will be discussed later. A typical such matrix is given by Volponi as

$$\left\{ \begin{array}{c} N1 \\ N2 \\ WF \\ T25 \\ T3 \\ T49 \\ P25 \\ PB \end{array} \right\} = \left[\begin{array}{ccccc} 0.28 & -0.37 & 0.11 & -0.24 & -0.01 \\ -0.03 & 0.05 & -0.15 & -0.04 & 0.23 \\ -0.25 & 0.34 & -0.12 & -0.48 & -0.83 \\ 0.26 & -0.36 & -0.23 & 0.28 & -0.24 \\ 0.02 & -0.02 & -0.26 & -0.01 & -0.51 \\ -0.19 & 0.26 & -0.29 & -0.28 & -1.02 \\ 0.82 & -1.13 & 0.17 & 0.95 & -0.79 \\ 0.02 & -0.03 & 0.02 & 0.00 & 0.22 \end{array} \right. $$

$$ \left. \begin{array}{ccccc} 0.00 & -0.01 & 0.01 & 0.46 & -0.25 \\ -0.29 & 0.35 & -0.10 & 0.00 & 0.20 \\ 0.05 & -1.11 & 0.28 & 0.02 & -0.63 \\ 0.01 & -0.33 & 0.08 & 0.38 & -0.39 \\ 0.00 & 0.15 & -0.31 & 0.09 & 0.02 \\ 0.06 & -1.36 & 0.34 & -0.60 & -0.44 \\ 0.04 & -1.09 & 0.26 & 1.18 & -1.24 \\ -0.05 & 0.50 & -1.05 & 0.25 & 0.07 \end{array} \right] \left\{ \begin{array}{c} FAN\ ETA \\ FAN\ FC \\ LPC\ ETA \\ LPC\ FC \\ HPC\ ETA \\ HPC\ FC \\ HPT\ ETA \\ A4 \\ LPT\ ETA \\ A49 \end{array} \right\}$$

An ideal solution for no sensor noise would be

$$\hat{x}_e = H_e^{-1}z \tag{1.24}$$

More realistic solutions are

$$\hat{x}_e = \left(H_e^T R^{-1} H_e\right)^{-1} H_e^T R^{-1} z$$

or   (1.25)

$$\hat{x}_e = P_0^T H_e^T \left(H_e P_0 H_e^T + R\right)^{-1} z$$

The influence coefficients can be expanded to include sensor faults. A typical sensor fault delta vector is given as

$$x_s = \left\{ \begin{array}{c} \Delta N1_{error} \\ \Delta N2_{error} \\ \Delta WF_{error} \\ \Delta T25_{error} \\ \Delta T3_{error} \\ \Delta T49_{error} \\ \Delta P25_{error} \\ \Delta PB_{error} \end{array} \right\}$$

The sensor influence coefficient matrix can then be written as

$$z = H_s x_s \tag{1.26}$$

Here the subscript s represents sensor error. In expanded form,

$$
\left\{ \begin{array}{c} \Delta N1_{error} \\ \Delta N2_{error} \\ \Delta WF_{error} \\ \Delta T25_{error} \\ \Delta T3_{error} \\ \Delta T49_{error} \\ \Delta P25_{error} \\ \Delta PB_{error} \end{array} \right\} =
\left[\begin{array}{ccccccccccc}
1 & 0 & 0 & 0 & 0 & 0 & 0 & 0 & 0.45 & -0.500 & -0.45 \\
0 & 1 & 0 & 0 & 0 & 0 & 0 & 0 & 0.20 & -0.500 & -0.20 \\
0 & 0 & 1 & 0 & 0 & 0 & 0 & 0 & 0.79 & -0.590 & -1.79 \\
0 & 0 & 0 & 1 & 0 & 0 & 0 & 0 & 0.22 & -0.985 & -0.22 \\
0 & 0 & 0 & 0 & 1 & 0 & 0 & 0 & 0.38 & -0.932 & -0.38 \\
0 & 0 & 0 & 0 & 0 & 1 & 0 & 0 & 0.39 & -1.000 & -0.39 \\
0 & 0 & 0 & 0 & 0 & 0 & 1 & 0 & -0.42 & 0.000 & -0.58 \\
0 & 0 & 0 & 0 & 0 & 0 & 0 & 1 & 0.48 & 0.000 & -1.48
\end{array} \right]
\left\{ \begin{array}{c} N1ERROR \\ N2ERROR \\ WFERROR \\ T25ERROR \\ T3ERROR \\ T49ERROR \\ P25ERROR \\ PBERROR \\ P2ERROR \\ T2ERROR \\ P49ERROR \end{array} \right\}
$$

The engine and sensor faults can then be combined to yield

$$z = H_e x_e + H_s x_s = \left[H_e \vdots H_s \right] \begin{bmatrix} x_e \\ \cdots \\ x_s \end{bmatrix} = Hx \qquad (1.27)$$

Typically, the engine/sensor influence coefficients H, measurement covariance matrix R, and measurement deltas z are known. The initial guess $\hat{x}_0$ and the state transition matrix φ need to be estimated as accurately as possible. The state covariance P_0 and process noise covariance Q are free parameters. Q is often set to zero.

1.9 Vibration-Based Diagnostics

Continuous vibration monitoring of aircraft engines during flight has become routine since the 1970s. The development of the piezoelectric accelerometer and use of trend monitoring was important for the deployment of such systems. Typically, the health of a rotating system such as a gas turbine is manifested by its vibration level. The amplitude of vibration at the (1/rev) frequency directly indicates the state of balance of the machinery. We have already seen that the state of the engine is indicated by measurements such as pressure and temperature. Also, debris monitoring, which involves the level of contamination in the oil system, is a good indicator of damage. However, pressure and temperature are point measurements and are thus local in nature. In contrast, vibration is a global phenomenon and contains information about the whole system. In fact, the problem is that raw vibration signals contain too much information about the system.

A typical vibration monitoring system must provide overall vibration and (1/rev) amplitude and phase data. A detailed spectral analysis can be often avoided by comparing (1/rev) data with the overall vibration level data. If the (1/rev) data and the overall vibration level have the same amplitude, then it can be concluded that the (1/rev) vibration constitutes the majority of the vibration signal and the other vibration components are much less in magnitude. On the other hand, if there is a large difference between (1/rev) data and the overall amplitude, it is time to do a detailed vibration analysis.

Typically, an accelerometer is placed at some points on the gas turbine, as shown in Figure 1.3. Here a gas turbine is shown schematically. It has two major modules: gas generator (GG) and the power turbine (PT). The gas generator is divided into five submodules: the compressor front frame (CFF),

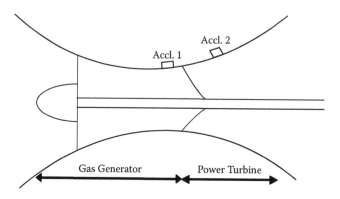

FIGURE 1.3
Accelerometer on the gas turbine.

the compressor rotor and stator, the compressor rear frame (CRF), the high-pressure turbine (HPT), and the turbine mid frame (TMF). The first accelerometer in Figure 1.3 is mounted on the CRF and the second is on the turbine rear frame (TRF).

The vibration data obtained from such accelerometers are stored and trended with time. Various diagnostic algorithms have been developed for vibration monitoring. Chapter 12 provides a case study of vibration monitoring approach for a turbine blade.

2

Idempotent Median Filters

Signal processing plays an important role in gas turbine diagnostics. In this chapter, we use a special type of median filter called the center weighted idempotent median (CWIM) filter to process gas turbine health signals for improved visualization and analysis. The filter requires only two forward data points. Thus, the CWIM filter is useful for gas turbines where data are available slowly. The filter is described in this chapter and then demonstrated on signals containing both deterioration and sharp trend shifts. A key advantage of this filter is that it does not need *a priori* knowledge of the noise characteristics of the signal. The idea of using the CWIM filter for gas turbine diagnostics was proposed by Ganguli [37] and is discussed in this chapter.

2.1 Weighted Median Filter

The simple median filter was introduced in Chapter 1 as a basic nonlinear filter that is good for removing non-Gaussian outliers. The weighted median (WM) filter is a generalization of the standard median filter where nonnegative integer weights are assigned to each position in the filter window. WM filters provide a number of free parameters in the form of weights that can be tuned to design a filter to perform specific tasks. WM filters have been successfully used in image processing where edges are very important details. The WM filter output is given as [19]

$$y(k) = median(w(k - K)^* \, x(k - K), \ldots w(k)^* \, x(k), \ldots w(k + K)^* \, x(k + K)) \quad (2.1)$$

where * stands for duplication. Duplication means that the sample $x(k)$ is repeated $w(k)$ times in the array before taking the median. For example, $3^*x(k)$ is the same as $x(k), x(k), x(k)$; i.e., the sample $x(k)$ is repeated three times. There are $N = 2K + 1$ weights for the WM filter. If we assume that Equation (2.1) is defined for the kth sample, we can write the definition of the WM filter in more compact form as

$$y = median(w_{-k}^* \, x_{-k}, \ldots, w_{-1}^* \, x_{-1}, w_0^* \, x_0, w_1^* \, x_1, \ldots w_k^* \, x_k) \quad (2.2)$$

Symmetric WM filters are widely used in nonlinear signal processing to avoid bias effects. A symmetric WM filter has weights satisfying the relationship $w_{-i} = w_i$, $i = 1, 2, ..., K$. It should be noted that WM filters with positive integer weights are limited to low-pass capabilities. Low-pass filters remove high-frequency noise. Arce and Parades [38] generalized the weights to negative weights by using the following definition:

$$y = median\left(|w_{-k}|^* \operatorname{sgn}(w_{-k})x_{-k}, \cdots, |w_{-1}|^* \operatorname{sgn}(w_{-1})x_{-1}, |w_0|^* \operatorname{sgn}(w_0)x_0,\right.$$

$$\left.|w_1|^* \operatorname{sgn}(w_1)x_1, \cdots, |w_k|^* \operatorname{sgn}(w_k)x_k\right) \tag{2.3}$$

Here the weight signs are uncoupled with the weight magnitudes and merged with observation samples. This extension to negative weights allows the use of the weighted recursive median (WRM) filter to do band-pass or high-pass filtering and suppress desired frequencies, respectively. The weights of the filter could be optimized for specialized application [38]. However, in our application to gas turbine diagnostics, we look for a low-pass filter contaminated with high-frequency Gaussian noise. Hence, we consider positive integer weights only. Finding appropriate weights for signals obtained from gas turbine engines is an issue that is addressed in this chapter in a simple manner. Later in this book, a more sophisticated approach is used to optimally obtain median filter weights.

2.2 Center Weighted Median Filter

A subclass of the symmetric weighted median filter is the center weighted median (CWM) filter. In the CWM filter, all samples inside the filter window are assigned unit weights except the center sample. Thus, a CWM filter with window size $2K + 1$ has a weight $w_0 = 2L + 1$ for the center sample, and all other weights are $w_i = 1$ for each $i \neq 0$, where L and K are nonnegative integers.

$$y = median (x_{-k}, ... x_{-1}, 2L + 1 * x_0, x_1, ... x_k) \tag{2.4}$$

Different CWM filters are produced by different values of L. When $L = 0$, the CWM filter reduces to the standard median filter, as all weights are equal to unity. When $L \geq K$, the CWM becomes the identity filter since the number of duplications of the center sample results in the median becoming equal to the center sample. Root structures for the CWM filter have the following theorems associated with them:

Theorem 2.1

The minimum length of a constant neighborhood of CWM of window length $2K + 1$ and center weight $2L + 1$ is $2K + 1 - L$. ∎

Theorem 2.2

An edge is a root of any CWM filter. ∎

Proofs of these theorems can be found in [22]. The second theorem shows that a step change in the signal is not disturbed by the CWM filter.

2.3 Center Weighted Idempotent Median Filter

When $L = K - 1$, the CWM filter is an *idempotent filter* [39], which produces root signals after a single filtering pass. This avoids the need for repeated passes that are needed by other types of median-based filters to converge to the root signal. Thus, a center weighted idempotent median (CWIM) filter of window length $2K + 1$ can be defined as

$$y = median(x_{-k}, \ldots x_{-1}, 2K - 1 * x_0, x_1, \ldots, x_k) \qquad (2.5)$$

2.3.1 Filter Design for Gas Path Measurements

The approach of designing a filter to preserve certain image details while discarding others is known as optimal filtering under structural constraints [22]. The CWM filter has two design parameters that must be determined in order to meet certain requirements. These are the center weight $2L + 1$ and the window length $2K + 1$. A typical design objective is to find a CWM filter to preserve certain signal structures, for example, the smallest length constant neighborhood, which is a set of adjacent points having similar values.

Using the theorems stated earlier, we could design a filter for specific applications. For gas path measurement deltas, we can assume that any one point that is not a part of a trend or constant neighborhood is a spurious data point representing impulsive noise. However, any two or more points that represent a trend are assumed to reflect a genuine trend. This is a very conservative assumption and ensures that any fine details in the image lasting over one point are preserved. We want a filter with minimum need for forward data. This can be accomplished by using a filter of small window length (for example, a three-point filter). However, larger window lengths lead to better noise attenuation. As a compromise, we select a five-point filter with window length $N = 2K + 1 = 5$. This results in $K = 2$ and a time delay of only two data points in the signal.

From Theorem 2.1, we see that the minimum length of a constant neighborhood of CWM of window length $2K + 1$ and center weight $2L + 1$ is $K + 1 - L$. Therefore, for preserving constant neighborhoods of minimum length 2, we need $K + 1 - L = 2$; this yields $L = 1$. The center weight is then equal to $2L + 1 = 3$. The corresponding filters with $K = 2$ and $L = 1$ are

$$y = median(x_{-2}, x_{-1}, 3 * x_0, x_1, x_2) \tag{2.6}$$

The above filter is also an idempotent filter since $L = K + 1$. Therefore, it should converge to a root signal in only one pass and not need repeated passes like the conventional median filters described in Chapter 1. We shall use the five-point CWIM filter with triple duplication of the center sample described by Equation (2.6) for our results in this chapter. The filter has the following desirable properties:

1. It preserves constant neighborhoods of minimum length equal to 2.
2. It does not preserve constant neighborhoods of length less than 2.

From Theorem 2.2, an edge is always preserved by the CWM filter. It therefore preserves fine details in the signal but removes spurious impulsive noise. Recall that the fine details in the signal are typically caused by single-fault events. Once the filter has been created, we test it for gas turbine applications.

2.4 Test Signal

Gas path measurement deltas are obtained by subtracting the baseline measurements for a good engine from the actual measurements. The baseline measurements often come from an engine model, and various correction factors are used to reduce the measured data to standard sea level conditions [40]. Since both the engine model and the correction factors are mathematical model idealizations, they are sources of errors in the gas path measurements deltas. Therefore, gas path measurement deltas contain high levels of uncertainty due to sensor errors and modeling approximations.

A typical twin spool gas turbine (Figure 1.2) consists of five modules: fan (FAN), low-pressure compressor (LPC), high-pressure compressor (HPC), high-pressure turbine (HPT), and low-pressure turbine (LPT). Air coming into the engine is compressed in the FAN, LPC, and HPC modules, combusted in the burner, and then expanded through the HPT and LPT modules producing power. The sensors *N1*, *N2*, *WF*, and *EGT* provide information about the condition of these modules and are used for diagnostics of gas turbine engines. In this chapter, we will use only these four measurements to analyze the filter.

Table 2.1 shows influence coefficients for a commercial gas turbine engine at a fixed power condition with $\eta = -2\%$ from a baseline good engine. These influence coefficients are taken from [12]. The numbers in this table are

TABLE 2.1

Fingerprints for Selected Gas Turbine Faults for $\eta = -2\%$

Faults	$\Delta EGT^\circ(C)$	$\Delta WF\%$	$\Delta N2\%$	$\Delta N1\%$
High-pressure compressor	13.6	1.6	−0.11	0.1
High-pressure turbine	21.77	2.58	−1.13	0.15
Low-pressure compressor	9.09	1.32	0.57	0.28
Low-pressure turbine	2.38	−1.92	1.27	−1.96
Fan	−7.72	−1.4	−0.59	1.35

Source: Ganguli, R., *Journal of Propulsion and Power* 19(5):930–937, 2003. With permission.

the fingerprints or fault signatures for the module faults. As an example, a 2% efficiency decrease in the high-pressure compressor corresponds to a 13.6°C increase in exhaust gas temperature, a 1.6% increase in the fuel flow, a 0.11% decrease in high rotor speed, and a 0.10% increase in low rotor speed. Test signals are created for the four measurements with the fingerprint chart numbers as a guide for the maximum measurement deltas. Using synthetic test signals allows us to evaluate the filter performance since the final answer is known.

2.4.1 Ideal Signal

The signals in Figures 2.1–2.4 contain 250 data points for the ΔEGT, ΔWF, $\Delta N2$, and $\Delta N1$ measurements, respectively. In each figure, an ideal, noisy, and CWIM FIR and IIR filtered signal are shown. The FIR and IIR filters have been discussed in Chapter 1. The ideal test signals used in this study are obtained by putting together linear deterioration signals superimposed with edges representing a single-fault or maintenance event. A number of combinations of the deterioration and fault signals are studied. This simulated fault time history idealizes a real-life scenario within a compressed timescale. The maximum amplitude values used for each signal are $\Delta EGT = 15.4$°C, $\Delta WF = 1.91\%$, $\Delta N2 = -0.74\%$, and $\Delta N1 = -1.99\%$, which corresponds to, for example, an HPT fault of $\eta = -1.41\%$, an HPT fault of $\eta = -1.48\%$, an HPT fault of $\eta = -1.31\%$, and an LPT fault of $\eta = -1.97\%$, respectively (from Table 2.1).

2.4.2 Noisy Signal

Random noise is added to the simulated measurements using standard deviations for ΔEGT, $\Delta N1$, $\Delta N2$, and ΔWF of 4.23°C, 0.25%, 0.17%, and 0.50%, respectively. These numbers are obtained from typical airline data given in [12]. Impulsive noise is also added to the ideal signal. The impulses are selected at eight levels: σ, 1.5σ, 1.75σ, and 2σ, and $-\sigma$, -1.5σ, -1.75σ, and -2σ. These points are placed in an arbitrary way to simulate spurious data that

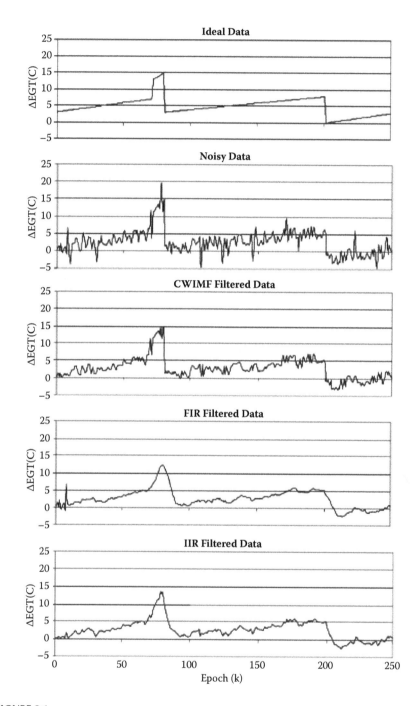

FIGURE 2.1
Ideal, noisy, and filtered data for deviations in exhaust gas temperature. (From Ganguli, R., *Journal of Propulsion and Power* 19(5):930–937, 2003. With permission.)

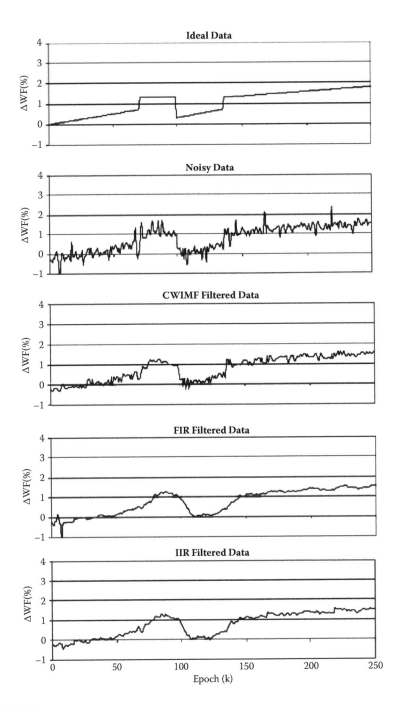

FIGURE 2.2
Ideal, noisy, and filtered fuel flow deviations. (From Ganguli, R., *Journal of Propulsion and Power* 19(5):930–937, 2003. With permission.)

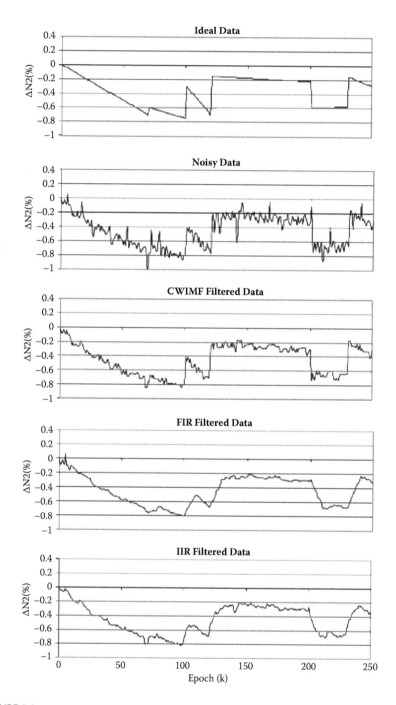

FIGURE 2.3
Ideal, noisy, and filtered deviations in *N2*. (From Ganguli, R., *Journal of Propulsion and Power* 19(5):930–937, 2003. With permission.)

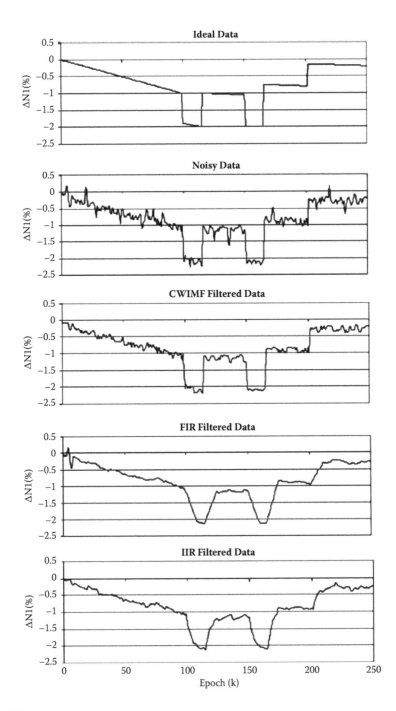

FIGURE 2.4
Ideal, noisy, and filtered deviations in $N1$. (From Ganguli, R., *Journal of Propulsion and Power* 19(5):930–937, 2003. With permission.)

follow no theoretical noise model. The noisy signal is shown in Figures 2.1–2.4 for the four measurements. Both random and impulsive noises are included in that signal. It can be observed that noise causes problems in differentiating between a healthy and damaged engine and also hides important features of the data.

After noise is added to this signal, it allows us to test the performance of a filter in the presence of trend shifts that can represent a single-fault precursor to a major maintenance event and also ramps, which models long-term deterioration over time. The stationary regions simulate a healthy engine. The test signals used here are relatively more complex than those found in actual practice. However, they serve to illustrate the CWIM, FIR, and IIR filters over a range of signal-to-noise ratios and for different deterioration rates and trend shifts.

2.5 Error Measure

Consider the basic measurement deltas ΔEGT, $\Delta N1$, $\Delta N2$, and ΔWF. We can write any of these measurement deltas as follows:

$$z = z^0 + \theta \tag{2.7}$$

where θ is noise and z^0 is the measurement delta, which is also called the ideal signal. In reality, such a pure signal would be contaminated by noise and outliers, and therefore z is the polluted or corrupted signal. A filter Ψ performs the following operation that returns the filtered signal from the corrupted signal:

$$\hat{z} = \Psi(z) = \Psi\left(z^0 + \theta\right) \tag{2.8}$$

In the next section, we evaluate the CWIM, FIR, and IIR filters using simulated data. The following root mean square error measure based on the L_2 norm will be used to analyze the filter performance over a sample of M points by comparing the filtered signal with the ideal signal:

$$\Theta = \frac{1}{M} \sum_{k=1}^{M} \sqrt{\left(\hat{z}_k - z_k^0\right)^2} \tag{2.9}$$

2.5.1 Numerical Simulations

Numerical studies are conducted to qualitatively and quantitatively evaluate the CWIM filter and the traditional linear filters using the test signals.

Figures 2.1–2.4 show the ideal, noisy, and filtered signals for ΔEGT, ΔWF, $\Delta N2$, and $\Delta N1$, respectively. It is verified that the ideal signal for each of the four cases is indeed a root signal of the CWIM filter. This is not surprising since constant regions, step edges, and ramp edges of sufficient extent are root signals of median filters, as mentioned by Senel et al. [21]. The FIR filter used in these figures uses a 10-point moving average (Equation (1.2)), and the IIR filter uses $a = 0.25$ in Equation (1.3).

The results also show that the CWIM filter works best in removing outliers and not in removing Gaussian noise. Thus, for the ΔEGT and ΔWF signals, which have high levels of Gaussian noise, the CWIM filtered signal still contains more random noise than the FIR and IIR filtered signals. For the $\Delta N2$ and $\Delta N1$ signals, which have low random noise levels, the filtered signal appears much less noisy. This is not surprising since the CWIM filter used here falls under the class of the "gentle filter," which removes outliers while not affecting other features to a large extent [19]. In contrast, the FIR and IIR filters remove Gaussian noise. However, fault isolation algorithms such as the Kalman filter, which is used for gas path state estimation, are only optimal under a Gaussian noise environment. We have already seen this in the discussion presented in Chapter 1, where added Gaussian noise was assumed in the measurements. Therefore, such algorithms can handle the Gaussian noise present in the data, but may have problems with non-Gaussian outliers.

It can be observed from Figures 2.1–2.4 that considerable noise is reduced for each signal after CWIM filtering, and that trend changes and deterioration history are more clearly visible. For the ΔEGT signal, in Figure 2.1, the three points where trend shifts occur are clearly identified. In addition, the linear variations simulating engine deterioration are also preserved. The impulsive outliers in the data are successfully removed. However, the FIR filter smoothes out the trend shifts. The FIR filter takes nine points to start, resulting in some impulsive noise in the signal between $k = 1$ and $k = 9$ not being smoothed. In contrast, the IIR filter needs one point to start and the CWIM filter needs two points. The IIR filter also reduces random noise but smoothes out the trend shifts in the signal. For the ΔWF signal in Figure 2.2, the three points where the trend shifts occur are clearly identified after CWIM filtering. The linear characteristics in the signal are also preserved.

For the $\Delta N2$ signal in Figure 2.3, the step edges are clearly preserved and their temporal locations enhanced in the signal after CWIM filters. In addition, the fine detail due to a small trend shift between $k = 50$ and $k = 100$ is also brought out in the filtered signal. Such fine detail is very difficult to decipher in the noisy signal. Finally, the $\Delta N1$ signal in Figure 2.4 shows that the step edges and deterioration features are clearly brought out. The results also show that the CWIM filter works best in removing outliers and not in removing Gaussian noise. Thus, for the ΔEGT and ΔWF signals, which have high levels of Gaussian noise, the CWIM filtered signal still contains more random noise than the FIR and IIR filtered signals. For the $\Delta N2$ and

$\Delta N1$ signals, which have low random noise levels, the filtered signal appears much less noisy. Recall that the CWIM filter falls under the class of the gentle filter, which removes outliers while not affecting other features to a large extent [19]. The impulses that corrupted the signal and prevented proper visualization have been removed by the CWIM filter. The removal of impulsive noise is important for improved visualization, as the human visual system is very sensitive to high frequency in the form of edges. The removal of the impulsive noise also makes the filtered signal more amenable to automated fault detection and isolation. The results shown in Figures 2.1–2.4 are qualitative and represent one of many possible noisy samples. For a more quantitative understanding, 1000 samples of noisy random data about the ideal signals shown in Figures 2.1–2.4 were taken for each of the four measurements and the average root mean square error calculated. These results are shown in Table 2.2. For each of the four measurements there is a reduction in noise of about 58–60% for the CWIM filtered signal compared to the noisy signal.

To illustrate the benefit of the CWIM filter over the linear filter, Figure 2.5 compares the noise reduction in ΔEGT using the FIR, IIR, and CWIM filters.

TABLE 2.2

Average Root Mean Square Error for Noisy and CWIM Filtered Data

	ΔEGT	ΔWF	$\Delta N2$	$\Delta N1$
Noisy	0.156	0.0046	0.0062	0.0091
Filtered	0.064	0.0019	0.0025	0.0037

Source: Ganguli, R., *Journal of Propulsion and Power* 19(5):930–937, 2003. With permission.

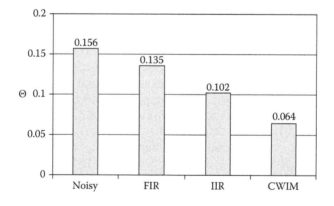

FIGURE 2.5
Average root mean squared error for noisy and FIR, IIR, and CWIM filtered ΔEGT data. (From Ganguli, R., *Journal of Propulsion and Power* 19(5):930–937, 2003. With permission.)

Compared to the noisy signal, the FIR filter shows a noise reduction of 13%, the IIR filter of 35%, and the CWIM filter of 59%. Therefore, the nonlinear CWIM filter can be recommended for noise removal from gas path measurements.

2.6 Summary

A nonlinear filter known as the center weighted idempotent median (CWIM) filter is analyzed for improved visualization and noise removal in gas turbine engine path measurements. A typical gas turbine engine measurement delta signal is created using linear deterioration superimposed with occasional trend shifts. The four measurements considered are exhaust gas temperature, fuel flow, low rotor speed, and high rotor speed. The CWIM filter is specially designed for noise removal in gas turbine engine measurement signals. This filter results in a noise reduction of about 60% in all four measurements used in this chapter. The CWIM filter retains the trend shifts and other features in the signal while removing noise. It helps in generating a signal that is more suited to the human visual system by removing high-amplitude impulsive noise that can lead to a person observing patterns where none are really present.

Filtering gas path measurements using the CWIM filter prior to fault detection and isolation are likely to improve the performance of state estimation algorithms such as the Kalman filter, which are optimal for Gaussian noise and can show performance degradation in the presence of non-Gaussian outliers. The linear FIR and IIR filters typically used for smoothing gas turbine engine signals are found to smooth out the key features in the signal. For the exhaust gas path temperature signal, the noise reductions by the FIR, IIR, and CWIM filters are 13, 35, and 59%, respectively. The CWIM filter is therefore recommended for preprocessing gas turbine engine measurement deltas before performing fault detection or visualization. The CWIM filter involves a very slight increase in complexity relative to the simple median filter, and is suited for engines where data are coming slowly. The following chapters will illustrate other nonlinear filters useful for gas turbine diagnostics.

3

Median-Rational Hybrid Filters

Diagnostic applications typically involve the detection and isolation of a system fault based on comparison between a good baseline system and a damaged system. Many diagnostic systems are designed based on mathematical models for the good and bad systems using methods that fall under the broad class of *model-based diagnostics*. A health signal can be interpreted as a measurement delta between the damaged measurement $z^{(d)}$ and undamaged measurement $z^{(u)}$ and written as $\Delta = z^{(d)} - z^{(u)}$. Under ideal conditions, the system has no fault, i.e., $\Delta = 0$. When a fault occurs, Δ assumes a nonzero value whose magnitude depends on the size and location of the fault. In this idealized system, the nonzero value of the measurement deviation, along with other measurement deviations, can be used to detect and isolate the fault. For commercial aircraft engines, only few data points are received for each flight. Therefore, it is important to keep the forward data point requirement to minimum. In Chapter 2, we looked at the "gentle" center weighted idempotent median (CWIM) filter for gas turbine diagnostics. In this chapter, we explore other filters with a low time delay for gas turbine applications.

Figure 3.1 shows a schematic of the gas turbine diagnostic process, which uses the engine measurement deltas to detect and isolate faults and then suggests prognostic action based on nondestructive testing, boroscope, and manual inspections of the fault module. It is clear that if the fault module is correctly identified, the cost of maintenance for the airline comes down.

The present chapter discusses the filter architecture shown in Figure 1.1, which is enclosed in the dotted rectangle in Figure 3.1. In this chapter, the median filter is used to remove non-Gaussian outliers and the rational filter is used to remove Gaussian noise. The hybridization of these filters and their application to gas turbine diagnostics were proposed by Verma and Ganguli [41] and are discussed in this chapter.

3.1 Test Signals

The gas path signal is modeled using (1) step and (2) ramp edges. Consider the time series for 25 points shown in the ideal step signal in Figure 3.2. This step signal simulates an abrupt fault. The onset of the fault is at discrete time $k = 12$. The noisy signal can be expressed as

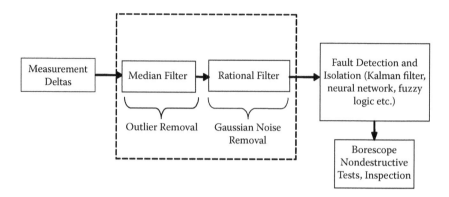

FIGURE 3.1
Schematic of noise and outlier removal in gas turbine diagnostics process. (From Verma, R., and Ganguli, R., *IEEE/ASME Transactions on Mechatronics* 10(4):461–464, 2005. With permission.)

$$\Delta = \Delta^0 + \alpha\varepsilon + \theta \qquad\qquad (3.1)$$

where Δ^0 is the pure signal, ε and θ are added Gaussian noise and outliers, respectively, and α is a parameter that allows control of the level of noise in the noisy health signal Δ.

Figure 3.2 shows the ideal step signal along with a noisy test signal with $\alpha = 0.2$. The outlier signal contains five points represented by $\theta = -1$ at $k = 7$, $\theta = 0.75$ at $k = 10$, $\theta = -0.75$ at $k = 14$, $\theta = 1$ at $k = 18$, and $\theta = -1.5$ at $k = 22$. These outliers are placed in an arbitrary way along the time series and do not follow any noise model. Lu et al. [11] calls these wild points, which tend to occur in gas path sensor measurements. Such wild points can lead to significant deterioration in the performance of fault detection and isolation algorithms and therefore need to be removed through a filtering process.

Figure 3.3 shows the ideal and noisy test signal for a ramp edge simulating engine deterioration. The outliers are placed at the same location as for the step signal. Table 3.1 shows the fingerprint chart for a large commercial engine similar to the United Technologies PW4000-94 engine in cruise condition with engine pressure ratio of 1.29 [25]. The fingerprints are fault signatures of the engine at a given steady flight condition and relate the faults in a given module to changes in the gas path measurements. For the fingerprints shown in Table 3.1, the measurement uncertainties for ΔEGT, ΔWF, $\Delta N2$, and $\Delta N1$ are 4.23°C, 0.50%, 0.17%, and 0.25%, respectively [25]. These values were obtained by a study of airline data and were also used for the CWIM filter in Chapter 2. Using these values for the four measurements, the signal-to-noise ratios are obtained by dividing the fingerprints in Table 3.1 with the corresponding measurement uncertainty. These results are shown in Table 3.2, where it can be seen that the signal-to-noise ratios

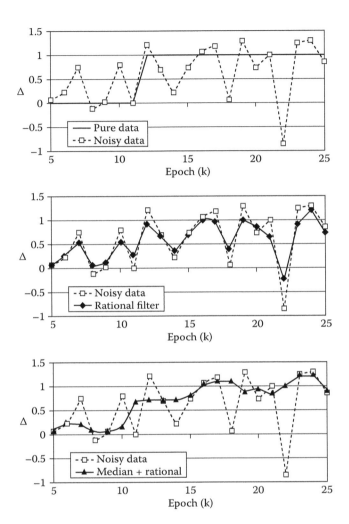

FIGURE 3.2
Ideal, noisy, and filtered signal for engine abrupt fault. (From Verma, R., and Ganguli, R., *IEEE/ASME Transactions on Mechatronics* 10(4):461–464, 2005. With permission.)

range from a low of 0.56 to a high of 7.84. Since the ideal test signals in Figures 3.2 and 3.3 have a maximum value of 1, a noise level of 0.10 leads to a signal-to-noise ratio of 10 and a noise level of 0.4 leads to a signal-to-noise ratio of 2.5.

Also note that the ideal signal varies from zero in the initial stage to values between 0 and 1 for the ramp edge in Figure 3.3. Therefore, wide ranges of signal-to-noise ratios are addressed using the variation in α from 0.1 to 0.4. Results shown later in this chapter will vary the noise level from 0.10 to 0.4 to allow for evaluation of the filter over a broad range of noise levels likely

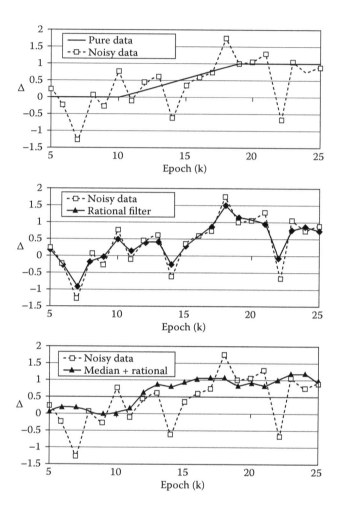

FIGURE 3.3
Ideal, noisy, and filtered signal for engine deterioration. (From Verma, R., and Ganguli, R., *IEEE/ASME Transactions on Mechatronics* 10(4):461–464, 2005. With permission.)

TABLE 3.1

Fingerprints for Selected Gas Turbine Faults for $\eta = -2\%$

Faults	ΔEGT	ΔWF	$\Delta N1$	$\Delta N2$
High-pressure compressor	13.6	1.6	−0.11	0.1
High-pressure turbine	21.77	2.58	−1.13	0.15
Low-pressure compressor	9.09	1.32	0.57	0.28
Low-pressure turbine	2.38	−1.92	1.27	−1.96
Fan	−7.72	−1.4	−0.59	1.35

Source: Verma, R., and Ganguli, R., *IEEE/ASME Transactions on Mechatronics* 10(4):461–464, 2005. With permission.

TABLE 3.2

Signal-to-Noise Ratios for Fingerprints of Gas Turbine Faults in Table 3.1

Faults	ΔEGT	ΔWF	$\Delta N1$	$\Delta N2$
High-pressure compressor	3.215	3.2	0.65	0.40
High-pressure turbine	5.15	5.16	6.65	0.60
Low-pressure compressor	2.14	2.64	3.35	1.12
Low-pressure turbine	0.56	3.84	7.47	7.84
Fan	1.82	2.80	3.47	5.40

Source: Verma, R., and Ganguli, R., *IEEE/ASME Transactions on Mechatronics* 10(4):461–464, 2005. With permission.

to occur in gas turbine applications. Note that the actual fault data for gas turbines are very difficult to obtain, and the use of simulated data allows the evaluation of the filter performance as the ideal signal is known. Furthermore, the types of signals used here have been used in the literature by Lu et al. [11] to evaluate a filtering approach based on autoassociative neural networks. The autoassociative neural networks and their use for signal denoising are discussed in Chapter 9.

Error measures are needed to quantify the noise removal capability of the filter. The mean square error (*MSE*) is a well-known measure that provides information about the filter accuracy and is defined as

$$MSE = \sum_{i=1}^{N} \frac{\left(\Delta - \Delta^0\right)^2}{N} \tag{3.2}$$

where N is the number of samples, Δ is the noisy or filtered signal value, and Δ^0 is the ideal or pure signal value. The noise reduction is defined in terms of percent as

$$\Theta^{(MSE)} = 100 \frac{MSE^{(noisy)} - MSE^{(filtered)}}{MSE^{(noisy)}} \tag{3.3}$$

The test signal and the error measure are used to evaluate the rational and median-rational filters.

3.2 Rational Filter

The filters used in this chapter are the rational filter and the median filter. The median filter was discussed in Chapter 1. Recall that the simple median filter used in this chapter is obtained by taking the median of a set of measurements. We now introduce the rational filter. The working of the rational filter is based on a nonlinear operator, which is able to attenuate the Gaussian noise in a signal, while preserving the edge to a good extent [42]. It is described by a rational function, which is defined in algebra as the ratio of two polynomials.

$$\hat{\Delta}_k = \frac{\Delta_{k-1} + \Delta_{k+1} + \Delta_k \left[\kappa(\Delta_{k-1} - \Delta_{k+1})^2 + \dfrac{1}{w} - 2 \right]}{\kappa(\Delta_{k-1} - \Delta_{k+1})^2 + \dfrac{1}{w}} \tag{3.4}$$

Here $\hat{\Delta}_k$ is the filtered signal value at discrete time k. The values of data at $k-1$, k, and kth time are Δ_{k-1}, Δ_k, and Δ_{k+1} and these points are the backward predictor, current value, and forward predictor, respectively. The parameters κ and w take positive values and are used to control the filter. The rational filter differs from the linear finite impulse response (FIR) filter mainly for the scaling, which is introduced on the Δ_{k-1} and Δ_{k+1} terms. Such terms are divided by a factor proportional to the edge sensing term. When $\kappa = 0$, the rational filter acts as the following linear filter:

$$\hat{\Delta}_k = w(\Delta_{k-1} + \Delta_{k+1}) + (1 - 2w)\Delta_k \tag{3.5}$$

The sum of the coefficients or weights of the above filter is 1. The filter shows low-pass behavior for $0 < w < 1/3$. For $w = 1/3$, the filter becomes a moving average filter. When $\kappa \to \infty$, the filter has no effect, and $\hat{\Delta}_k \cong \Delta_k$. For intermediate values of κ, the $(\Delta_{k-1} - \Delta_{k+1})^2$ term perceives the presence of a detail and accordingly reduces the smoothing effect of the operator. This filter has good edge preserving capability, which is required for gas turbine diagnostic problems. It has a time delay of only one point due to the forward predictor point Δ_{k+1}.

3.3 Median-Rational Filter

We now turn to the median filter, which belongs to a group of nonlinear filters called order statistics filters, which are based on sorting of the signal sample. The order statistics filters select one of the sorted neighborhood samples of the input signal vector in each sampling period. The three-point median filter can be written as

$$\hat{\Delta}_k = median\left(\Delta_{k-1}, \Delta_k, \Delta_{k+1}\right) \tag{3.6}$$

The median filter shown in Equation (3.6) has a one-point time delay and uses a forward and backward predictor. As mentioned in Chapter 1, the median filter is widely used in signal and image processing for the capability of outlier removal while preserving edges. However, the median is a selection filter, which means that its output is limited to one of the input samples. Therefore, the median is not very good at removing Gaussian or random noise since each element of the input sample contains random noise. However, the median filter is conceptually very simple, though long-length median filters involve sorting operations that can be computationally expensive. Therefore, in this chapter, we use only a three-point median filter to avoid forward point requirements and keep the computational expenses down. Note that the CWIM filter discussed in Chapter 2 has a two-point time delay.

This median-rational filter preprocesses the signal with a median filter before using the rational filter. We propose and use this combination in this chapter for denoising of gas path measurement deltas. First, the measurement delta is passed through the median filter. During this phase, the outliers in the data are removed. In the next phase, the median preprocessed data are sent through the rational filter. The rational filter can be defined using the outputs of the median filter from Equation (3.6) as

$$y_{k-1} = median(\Delta_{k-2}, \Delta_{k-1}, \Delta_k)$$

$$y_k = median(\Delta_{k-1}, \Delta_k, \Delta_{k+1}) \tag{3.7}$$

$$y_{k+1} = median(\Delta_k, \Delta_{k+1}, \Delta_{k+2})$$

The evaluation of the filtered value using the above formulas is very fast if one defines the denominator of the rational filter as a variable d_k and then uses it for the calculation of the median plus rational filter output as shown below:

$$d_k = \kappa(y_{k-1} - y_{k+1})^2 + \frac{1}{w} \tag{3.8}$$

$$\hat{\Delta}_k = \frac{y_{k-1} + y_{k+1} + y_k[d_k - 2]}{d_k} \tag{3.9}$$

The three-point median involves a simple sorting operation of three numbers and then taking the middle value. Therefore, the median plus rational approach is computationally efficient with a two-point time delay requiring points within a five-point window, including the $k-2$, $k-1$, k, $k+1$, and $k+2$ discrete time points. In this filter, κ has been fixed at 0.01 and w at 0.16, as suggested by Ramponi, who obtained these values as optimal for a signal contaminated with Gaussian noise [42]. This is a good assumption for the median plus rational approach as the data are first subjected to a median filter that removes outliers, and the rational filter is then used on a signal with trend shift and Gaussian noise.

3.4 Numerical Simulations

Simulations are performed using the test signal shown in Figures 3.2 and 3.3, which contain the step signal and the ramp signal simulating abrupt fault and engine deterioration, respectively. The noisy data shown in these figures use $\alpha = 0.2$ and added outliers that were discussed earlier. Figures 3.2 and 3.3 also show the results of processing the noisy signal using the rational and median plus rational filters, respectively.

From Figure 3.2, we see that the rational filter is able to preserve the trend shift while reducing Gaussian noise to some extent, but is unable to discard the outliers. The median plus rational approach results in the outliers being removed from the signal and Gaussian noise being reduced while preserving the trend shift. Figure 3.3 shows that for a linear signal, the rational filter is unable to remove outliers. However, the median plus rational approach results in outlier removal along with removal of some Gaussian noise. The visual quality of the signals is considerably improved after the application of the median plus rational filter. Since monitoring trend plots of gas path measurement deltas form an important diagnostic tool for airline power plant engineers, the use of the nonlinear filters for smoothing the data can greatly increase their capability of finding faults by visual inspections of the gas path sensor data themselves.

The above results are qualitative and provide visual information about the filters. However, they represent only one of many possible noisy data samples. To obtain quantitative results, 1000 samples of noisy data are created about the ideal step and ramp signals in Figures 3.2 and 3.3, and the noise reduction after filtering is calculated. Tables 3.3 and 3.4 show the noise reduction based on the mean square error (*MSE*) for the step signal and the ramp signal, respectively. The noise level added to the ideal signal varies

TABLE 3.3

Noise Reduction (%) with Filters for Engine Abrupt Fault Signal Based on *MSE*

α	0.1	0.15	0.2	0.25	0.3	0.35	0.4
Rational	46.93	46.95	46.99	47.05	47.13	47.21	47.29
Median	86.39	83.21	79.41	75.36	71.43	67.94	65.04
Median + rational	88.37	85.96	83.11	80.11	77.23	74.68	72.58

Source: Verma, R., and Ganguli, R., *IEEE/ASME Transactions on Mechatronics* 10(4):461–464, 2005. With permission.

TABLE 3.4

Noise Reduction (%) with Filters for Engine Deterioration Signal Based on *MSE*

α	0.1	0.15	0.2	0.25	0.3	0.35	0.4
Rational	48.44	48.39	48.34	48.3	48.27	48.25	48.23
Median	95.38	91.56	86.91	82	77.31	73.13	69.6
Median + rational	96.18	93.38	89.96	86.33	82.85	79.74	77.11

from low ($\alpha = 0.10$) to high ($\alpha = 0.40$). The rational filter reduces noise by 47% for the step signal and about 48% for the ramp signal. The noise reduction is almost constant across the noise levels. The median filter reduces the noise level by 65–86% for the step signal and 70–96% for the ramp signal. For the median filter, the noise reduction decreases with increasing levels of Gaussian noise in the data. The median filter works better when the data have outliers and low levels of Gaussian noise. The median plus rational filter reduces noise by 73–88% for the step signal and 77–96% for the ramp signal. The median plus rational filter gives more noise reduction than the median filter at all noise levels. However, the advantage increases at higher levels of Gaussian noise where the random noise-removing ability of the rational filter is useful.

The above results show that for gas path measurement data, which often contain high levels of outliers and random noise, it is advantageous to use the median plus rational filter to preprocess the data before performing fault detection and isolation functions. This filter is useful for jet engines where data are obtained very slowly, for example, at one or two points per flight.

3.5 Summary

Simulated diagnostic test signals are used to evaluate the denoising capability of nonlinear filters for smoothing gas turbine health signals. Linear filters such as the moving average, which are widely used in the gas turbine industry, tend to smooth out the sharp edges in the signal, which is often a precursor to an abrupt fault. Linear filters are also not good at removing outliers. The effect of both Gaussian noise and outliers of non-Gaussian origin are considered. The nonlinear filters used in this chapter are the rational filter and median filter.

The median and rational filters show good edge preservation capability. However, the rational filter is not good for outlier removal though it preserves the edges in the health signal. If the data are preprocessed by a median filter and then sent through a rational filter, both outliers and Gaussian noise are removed while preserving the edges in the signal, which are often precursors to abrupt faults.

The median plus rational filter results in noise reduction of 73–96% for the noisy signals and is also conceptually simple and computationally efficient when implemented in a small window of three points. Furthermore, the filter has a two-point time delay, making it suitable for gas turbine diagnostics where few points are obtained for each flight and the cost of transmitting additional points is high. The median plus rational filter is therefore recommended for preprocessing gas turbine measurement deltas before performing fault detection and isolation functions.

4

FIR-Median Hybrid Filters

We have seen in the previous chapter that linear filters can smooth out sharp edges in the signal that may indicate the onset of a single fault. Nonlinear filters such as median filters are able to preserve edges while simultaneously reducing noise and handle non-Gaussian noise such as outliers. Median filters are a well-known class of nonlinear filters in the image processing field. Median filters suffer from some shortcomings that can be alleviated by FIR-median hybrid (FMH) filters that combine the noise removal ability of FIR filters with the edge-preserving and outlier-removing ability of median filters. The FMH filter has been recently applied to the analysis of biomedical signals. In this chapter, a special FMH filter tuned to linear and step variations in a signal (typical of gas turbine health signals) is identified and shown to be superior to the currently used linear filters. In contrast to the filters discussed in Chapters 2 and 3, the FMH filter in this chapter has a long time delay and is suitable for engines where data are available at high speeds, such as for online health monitoring systems that can get real-time data. The application of FMH filters for gas turbine diagnostics was suggested by Ganguli [43] and forms the basis of this chapter.

4.1 FIR-Median Hybrid (FMH) Filters

FMH filters combine the desirable properties of the FIR filter for noise removal in constant signals and the capability of median filters to preserve edges, and can be written as [44]

$$\hat{x}_k = median\left(\Phi_1(x_k), \Phi_2(x_k), \cdots, \Phi_M(x_k)\right) \qquad (4.1)$$

where Φ_i is the ith FIR subfilter operating on the input signal. The length k and number M of FIR subfilters are selected to allow for an acceptable compromise between noise reduction and edge preservation. Typically, M is kept small in Equation (4.1) to allow for simple implementation. The large number of compare and swap operations needed by the median filter is eliminated in the FMH filter by taking the median over the outputs of a few FIR substructures. FMH filters are much faster than median filters of the same length. The median filter is a special case of the FMH filter where the lengths of all the FIR substructures are 1.

4.2 Weighted FMH Filter

By weighting the different FIR substructures, the resulting FMH filter can be tailored for different problems. The general weighted FMH filter has the form [48]

$$\hat{x}_k = median\left(w_1 * \Phi_1(x_k), w_2 * \Phi_2(x_k), \cdots, w_M *(x_k)\right) \qquad (4.2)$$

where $*$ denotes duplication, $w_i \geq 0$, and Σw_i is odd. There are many possibilities for the weights of the filter, and the actual weights depend on the application in question. The two extremes in weightings range from all the weights being equal to 1 (no duplication) to the idempotent filters that produce the root signal from noisy input in one filter pass.

A popular choice of number of subfilters is five, with the center filter being the current point itself. A five-point weighted FMH filter structure with weight symmetry with respect to the center point of the window of the filter is used in this study. This filter structure was chosen because it is general enough to be uniformly applicable and not too complicated to be implemented or analyzed [44].

The two types of subfilters mainly used as substructures are the average and the ramp predictors that are zero- and first-order FIR filters, respectively. The FIR subfilters characterize the root signals of FMH filters. For example, FMH filters with only mean filters as substructures have root signals that consist of steps and constant levels. When FMH filters with first-order substructures are used, root signals involving linear variations such as ramps can be considered. Higher-order polynomials can also be used to create FIR subfilters for more complex signals.

The weighted FMH filter with length $2I+1$ and five FIR substructures can be written as

$$\hat{x}_k = median\left(w_1 * y_1, w_2 * y_2, w_3 * y_3, w_4 * y_4, w_5 * y_5\right) \qquad (4.3)$$

where

$$y_1 = \frac{1}{I}(x_{k-1} + x_{k-2} + x_{k-3} + \cdots + x_{k-I})$$

$$y_2 = h_1 x_{k-1} + h_2 x_{k-2} + h_3 x_{k-3} + \cdots + h_I x_{k-I}$$

$$y_3 = x_k$$

$$y_4 = \frac{1}{I}(x_{k+1} + x_{k+2} + x_{k+3} + \cdots + x_{k+I})$$

$$y_5 = h_1 x_{k+1} + h_2 x_{k+2} + h_3 x_{k+3} + \cdots + h_I x_{k+I}$$

The current point is contained in y_3, y_1 is the backward zero-order FIR predictor, y_2 is the backward first-order FIR predictor, y_4 is the forward zero-order predictor, and y_5 is the forward first-order predictor. The predictive subfilters can be designed in a way such that they act as optimal predictors for an ith-order polynomial signal corrupted by Gaussian white noise. The weights for the optimal first-order filter of length I can be written as

$$h_i = \frac{4I - 6i + 2}{I(I-1)} \tag{4.4}$$

The filter discussed above has a very large number of possibilities depending on the weights selected. However, as mentioned in [48], it is not possible to analyze all the possible weighted filters, given the very large number of possibilities. The basic idea behind selecting useful weights is to amplify the importance of some of the subfilters in the median operation. Thus, if the center point is given more weight, the current data point is given more importance in the median, resulting in an increase in the number of signals the filter can handle at the cost of noise removal. If the zero-order subfilters are given more weights, the edges are preserved at the cost of restricting the root signals of the filter.

Two filter structures are interesting, one the center weighted FMH (CWFMH) filter and the other the subfilter weighted FMH (SWFMH) filter. These two filters are analyzed in detail in [44]. It is found that the CWFMH filter ($w_1 = 1$, $w_2 = 1$, $w_3 = 3$, $w_4 = 1$, $w_5 = 1$) leaves edges intact and preserves sinusoidal signals, so it is useful for continuous-wave signals. The SWFMH filter with weights ($w_1 = 2$, $w_2 = 1$, $w_3 = 1$, $w_4 = 2$, $w_5 = 1$) preserves edges while removing high-frequency noise from stationary signals.

The SWFMH filter discussed above is very well suited for gas turbine health signals that have sharp edges and linear components. It is compared to the traditional smoothing approaches based on FIR (averaging) filters and IIR (exponential) filters, along with the standard median filter. The SWFMH has been used for removing noise from biomedical EEG signals and EOG signals while preserving features. In this chapter, we will demonstrate the application of the SWFMH filter for gas turbine signals.

4.3 Test Signal

The test signal used in this study is obtained using a linearized model of a gas turbine at a given power condition obtained from published literature [12]. An ideal root signal for ΔEGT with implanted high-pressure compressor (HPC) and high-pressure turbine (HPT) faults is used in this study and is shown in Figure 4.1. A known root signal is a standard method for

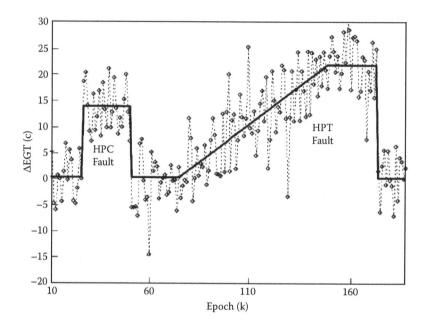

FIGURE 4.1

Root and noisy signal for exhaust gas temperature measurement delta indicating an HPC fault and HPT fault. (From Ganguli, R., *Mechanical Systems and Signal Processing* 16(6):967–978, 2002. With permission.)

evaluating different filters since the final answer is known. The other root signals for $\Delta N2$, $\Delta N1$, and ΔWF can be similarly derived from the engine model. However, to test the filters, we use the ΔEGT signal.

4.3.1 Root Signal

The signal in Figure 4.1 contains 200 data points. From point $k = 1$–25 there is no fault, representing a healthy engine. At data point $k = 26$, there is a sudden onset of a $\eta = -2\%$ HPC fault. Such an event can be triggered by an event such as foreign object damage and is a single fault. The root cause for this fault is identified and the HPC module repaired at point $k = 50$. However, after some time of fault-free operation, an HPT fault starts developing due to deterioration. However, unlike the HPC fault, which occurred suddenly, the HPT fault develops slowly over time (approximated by a linear function) from points $k = 75$–150; from $k = 150$ the HPT fault remains steady at $\eta = -2\%$ and is finally repaired at $k = 175$, after which normal operation resumes.

This simulated fault time history idealizes a real-life scenario within a compressed timescale. In the ideal scenario, the measurement delta is clearly defined at each point. After noise is added to this signal, it allows us to test the performance of the filter in the presence of edges that can represent a single-fault precursor to a major maintenance event and also a ramp, which models long-term deterioration over time. The stationary regions simulate a healthy engine.

4.3.2 Gaussian Noise

Zero mean Gaussian white noise is added to the simulated measurements using standard deviations for ΔEGT, $\Delta N1$, $\Delta N2$, and ΔWF of 4.23°C, 0.25%, 0.17%, and 0.50%, respectively. These values are obtained from typical airline data given in [11] and have been used in earlier chapters. The noisy signal is shown along with the root signal for ΔEGT in Figure 4.1. It is clear that noise causes problems in differentiating between a healthy and damaged engine and also hides important features of the data.

4.3.3 Outliers

Outliers are added to the root signal. The outliers are selected at three levels: σ, 2σ, and 3σ. The σ outlier is equal to 4.23°C and is added at $k = 10$, 80, and 140 and subtracted at $k = 40$ and 120; the 2σ outlier is equal to 8.46°C and is added at $k = 20$, 100, and 190 and subtracted at $k = 30$ and 170; and the 3σ outlier is equal to 12.69°C and is added at $k = 110$ and 160 and subtracted at $k = 60$ and 130. These points are selected in an arbitrary way to simulate gross outliers that follow no noise model. They are directly added to the root signal.

4.3.4 Error Measure

Consider the basic measurement deltas ΔEGT, $\Delta N1$, $\Delta N2$, and ΔWF. Then we can write any of these measurement deltas as follows:

$$z = z^0 + \varepsilon + \theta \tag{4.5}$$

Here ε accounts for Gaussian noise, θ represents non-Gaussian outliers, and z^0 is the pure measurement delta, also called the root signal. In reality, such a pure signal would be contaminated by noise and outliers, and therefore z is the polluted or corrupted signal. A filter Ψ performs the following operation that returns the filtered signal from the corrupted signal:

$$\hat{z} = \Psi(z) = \Psi\left(z^0 + \varepsilon + \theta\right) \tag{4.6}$$

In the next section, we evaluate numerical results using the simulated data for the FIR moving average filter, the IIR exponential average filter, the median filter, and the nonlinear SWFMH filter. The following *RMS* error measure is used to analyze the filter performance over a sample of N points by comparing the filtered signal with the root signal:

$$\Theta = \frac{1}{N}\sqrt{\sum_{k=1}^{N}\left(\hat{z}_k - z_k^0\right)^2} \tag{4.7}$$

4.4 Numerical Simulations

The traditional FIR and IIR filters are compared with the median and SWFMH filters. The FIR filter used in this chapter has window length $I=10$ and equal weights. It is therefore equivalent to the 10-point moving average that is commonly used in industry. The IIR filter used in this chapter is the exponential filter with $a=0.15$, which is one of the exponential smoothing filters suggested in [6]. The SWFMH filter used here has been discussed earlier in Equation (4.3). For applications with the median SWFMH filter, it is best to start with a small filter window and slowly increase the filter window. This helps to preserve features and reduce noise [19]. For the median filter, we start with a length of $I=3$, and then increase the filter length to $I=5, 7$, and 9. Note that the median filter must have an odd length. The filter length of the five underlying FIR structures of the SWFMH filter is started at $I=2$ and increased in increments of 1 to $I=10$. Thus, the SWFMH filter starts with a window length of $2I+1=5$ for $I=2$ and ends with a window length of $2I+1=21$ for $I=10$.

The results of filtering on the root signal embedded with Gaussian noise and the outliers are shown in Figures 4.2–4.5 for the FIR, IIR, median,

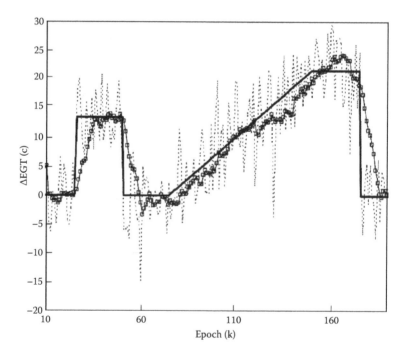

FIGURE 4.2
Effect of FIR filters on noisy signal contaminated with outliers. ━━━, root signal; •••••••, noisy signal; —□—, filtered signal. (From Ganguli, R., *Mechanical Systems and Signal Processing* 16(6):967–978, 2002. With permission.)

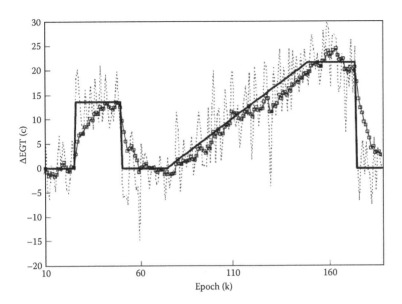

FIGURE 4.3
Effect of IIR filters on noisy signal contaminated with outliers. (From Ganguli, R., *Mechanical Systems and Signal Processing* 16(6):967–978, 2002. With permission.)

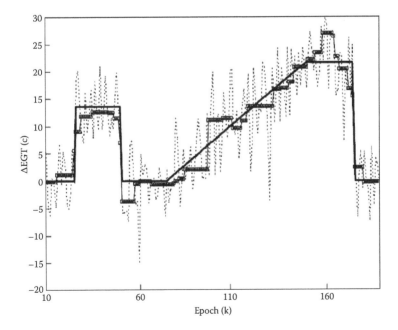

FIGURE 4.4
Effect of median filter on noisy signal contaminated with outliers. (From Ganguli, R., *Mechanical Systems and Signal Processing* 16(6):967–978, 2002. With permission.)

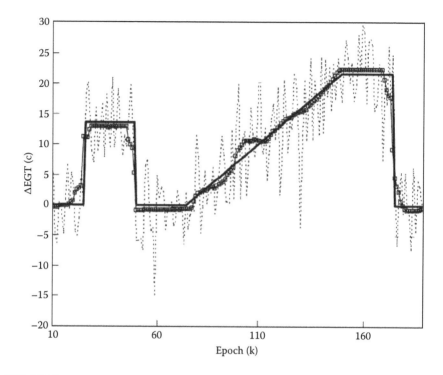

FIGURE 4.5
Effect of SWFMH filters on noisy signal contaminated with outliers. (From Ganguli, R., *Mechanical Systems and Signal Processing* 16(6):967–978, 2002. With permission.)

and SWFMH filters, respectively. The linear FIR and IIR filters are applied only once, as is the practice in industry, as repeated applications will smooth out features in the signal. The linear filters reduce Gaussian noise but also distort the edges in the signal. The linear filters are unable to reject the outliers, which get absorbed into the filter predictions.

In contrast, the median filter does a good job at preserving the signal features and effectively eliminates outliers because of the nonlinear nature of the median operation. However, it suffers from the problem of streaking, resulting in the horizontal lines approximating parts of the signals, especially along the ramp. The SWFMH filter is able to reproduce the root signal to a large extent from the noisy data. It effectively eliminates the outliers and preserves the sharp edges and the ramp features of the signal. The streaking problem of the median filter is also greatly diminished.

The above results are shown for one particular noisy signal. To get a statistical measure of the effectiveness of each filtering method, 1000 samples of noisy data with outliers are generated and filtered in a manner discussed above. The average *RMS* error defined by Equation (4.7) is calculated for each filter and normalized by the average *RMS* error for the noisy signal. These results are shown in Figure 4.6.

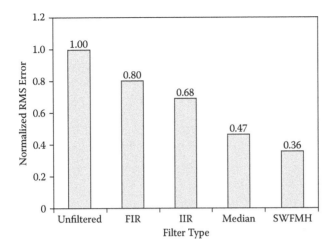

FIGURE 4.6
Filtering errors for different filters for noisy signal contaminated with outliers, normalized by error in noisy signal contaminated by outliers. (From Ganguli, R., *Mechanical Systems and Signal Processing* 16(6):967–978, 2002. With permission.)

Compared to the noisy signal, the FIR filter shows a noise reduction of 20%, the IIR filter of 32%, the median filter of 53%, and the SWFMH filter of 64%. Thus, the SWFMH filter gives excellent results in reducing noise in the signal while preserving the step changes and linear deterioration that are typical gas turbine health signals. The SWFMH filter is considerably superior to the widely used FIR filters and IIR filters [6]. Their use in the gas turbine industry for removing noise and outliers from health signals is therefore recommended.

4.5 Summary

The FIR-median hybrid filters are evaluated for removing noise and outliers from gas turbine measurement deltas. The measurement deltas are the deviations between the sensor measurements and a good baseline engine. A test signal derived from a linearized model of an engine at a fixed power condition is used to evaluate the FIR, IIR, median, and subfilter weighted FIR-median hybrid (SWFMH) filter.

It is found that the weighted median FIR hybrid filters provide an effective way to remove both Gaussian noise and outliers from measurement deltas, while preserving the edges in the signal. They are also able to capture the linear change in signals that can come from long-term engine deterioration.

The FIR, IIR, median, and SWFMH filters result in a noise reduction of 20, 32, 53, and 64%, respectively. The SWFMH filter is therefore recommended for noise removal in gas turbine health signals. It is particularly useful for cases where the data points are rapidly available, which is the case for many newer engines. They can also be used for online condition monitoring systems where real-time analysis is possible.

5

Transient Data and the Myriad Filter

In this chapter, we look at the myriad filter as a substitute for the moving average filter, which is widely used in the gas turbine industry. In contrast to previous chapters, which used steady-state signals, the current chapter considers transient signals. Typically, most gas turbine diagnostics are conducted with steady-state measurement data. Some gas turbine problems such as misscheduled nozzle and compressor blade movement due to control system faults appear only during transient processes [45]. The three ideal test signals used in this study include the step signal, which simulates a single fault in the gas turbine, while ramp and quadratic signals simulate long-term deterioration. Further, an adaptive weighted myriad filter algorithm that adapts to the quality of incoming data is studied. The filters are demonstrated on clean and simulated deteriorated engine data obtained from an acceleration process from idle to maximum thrust condition. These data were simulated using a transient performance prediction code. The deteriorated engine had single-component faults in the low-pressure turbine (LPT) and intermediate-pressure compressor (IPC). The signals are obtained from $T2$ (IPC total outlet temperature) and $T6$ (LPT total outlet temperature) engine sensors with their nonrepeatability values, which were used as noise levels. The weighted myriad filter shows greater noise reduction and outlier removal when compared to the sample myriad and FIR filter in the gas turbine diagnosis. Adaptive filters such as those considered in this chapter are also useful for online health monitoring, as they can adapt to changes in quality of incoming data.

5.1 Steady-State and Transient Signals

The test signals used in this chapter are for transient data. It is possible to use both steady-state and transient data for diagnostics. However, good steady-state data may not be available under all operating conditions. For example, military engines can operate up to 70% of the time in unsteady conditions [46]. In addition, some faults may be amplified under transient conditions. Several authors have looked at engine diagnostics with transient data [46–49]. Some studies address the problem of fault detection, and a few have looked at assessment of the fault magnitude. Neural networks and genetic algorithms have been used for the fault isolation problem. While the neural approach is robust to the presence of some level of noise in data, all fault isolation algorithms benefit from prior signal processing. Since transient data are acquired online at a reasonably high rate, we study an online adaptive signal processing method in this chapter, which can be useful for data cleaning prior to fault isolation and visual display to human engineers. Simulation of engine performance in real-time using a piecewise linear-state variable engine model (SVM) was used along with a Kalman filter to estimate engine performance [47]. This real-time Kalman filter can detect engine faults such as hardware failure, foreign object damage, etc. Such real-time systems can benefit if their signals are preprocessed by the myriad filter discussed in this chapter.

Linear filtering algorithms used in practical applications are limited to the cases of Gaussian noise and show performance degradation in the presence of impulsive contamination. This issue has been adequately discussed in the previous chapters. The class of myriad filters has been proposed for filtering highly noisy data [50–52]. In this chapter, we look at the myriad filter as a substitute for the moving average filter, which is widely used in the gas turbine industry. The idea of using the myriad filter for gas turbine diagnostics was first proposed by Surender and Ganguli [53] and is discussed in this chapter.

5.2 Myriad Filter

Myriad filters are nonlinear filters whose development was motivated by the properties of α-stable distributions, a family of heavy-tailed densities that have been proposed for robust non-Gaussian signal processing in impulsive noise environments [16]. Using the α-stable noise model leads to a noise removal processor that performs nonlinear operations on the data and is good at outlier removal. Myriad filters have a solid theoretical basis, and are inherently more powerful than median filters. However, they are also more complicated.

The myriad filter is defined as a running-window filter outputting the sample myriad of the elements in the window. In the case that all the weights are unitary [16], this filter is referred to as the unweighted myriad filter, or simply the myriad filter. The myriad filter is then defined as

$$\hat{\beta}_K = \arg\left(\min_{\beta} \prod_{i=1}^{N} \left[K^2 + (x_i - \beta)^2 \right] \right) \tag{5.1}$$

The myriad is the argument β that minimizes the product expression in the above equation. The myriad filters can be controlled by adjusting the linearity parameter K [51]. The larger the value of K, the closer the behavior of the myriad is to a linear estimator. As the myriad moves away from the linear region (large values of K) to lower values, the estimator becomes more nonlinear and resistant to the presence of impulsive noise.

We use a five-point myriad filter that represents a small window. Distortion to sharp edges in the signal that can precede single fault is minimal. This filter can be evaluated by using the following polynomial, which is obtained from Equation (5.1) by using $N = 5$:

$$f(\beta) = \left(K^2 + (x_1 - \beta)^2\right) * \left(K^2 + (x_2 - \beta)^2\right) * \left(K^2 + (x_3 - \beta)^2\right)$$
$$* \left(K^2 + (x_4 - \beta)^2\right) * \left(K^2 + (x_5 - \beta)^2\right) \tag{5.2}$$

The myriad is the value β that minimizes $f(\beta)$, where it lies between $[x_1, x_5]$, where x_1 and x_5 are the minimum and maximum values in the sample. We see from Equation (5.2) that $f(\beta)$ for the five-point myriad filter is a polynomial in β of order 10. In general, $f(\beta)$ for an N-point myriad is a polynomial of order $2N$. To find the value of the myriad, we need to find the roots of $f'(\beta)$ and the value of β among the root values that give the lowest $f(\beta)$. In this chapter the myriad filters are compared to finite impulse response (FIR) moving average filters of equal length. The FIR filter is defined as

$$\hat{x} = \sum_{i=1}^{N} w_i x_i \tag{5.3}$$

where the weights sum to 1. When the weights are equal, a moving average filter is obtained. The five-point moving average FIR filter is defined as

$$\hat{x} = \frac{x_1 + x_2 + x_3 + x_4 + x_5}{5} \tag{5.4}$$

where each weight is equal to 1/5. Note that the myriad filter becomes the moving average as $K \to \infty$. Defining $f_i(\beta) = (K^2 + (x_i - \beta)^2)$, Equation (5.2) can be written as

$$f(\beta) = \prod_{i=1}^{5} f_i(\beta) \tag{5.5}$$

Then

$$f'(\beta) = f_1' f_2 f_3 f_4 f_5 + f_1 f_2' f_3 f_4 f_5 + \cdots + f_1 f_2 f_3 f_4 f_5' = 0 \tag{5.6}$$

Dividing by $f_1 f_2 f_3 f_4 f_5$ gives

$$\frac{f_1'}{f_1} + \frac{f_2'}{f_2} + \cdots + \frac{f_5'}{f_5} = 0 \tag{5.7}$$

where $f_i' = -2(x_i - \beta)$. Now Equation (5.7) can be written as a moving average FIR filter.

$$\frac{x_1 - \beta}{1 + \dfrac{(x_1 - \beta)^2}{K^2}} + \frac{x_2 - \beta}{1 + \dfrac{(x_2 - \beta)^2}{K^2}} + \cdots + \frac{x_5 - \beta}{1 + \dfrac{(x_5 - \beta)^2}{K^2}} = 0 \tag{5.8}$$

As $K \to \infty$, $\dfrac{(x_i - \beta)^2}{K^2} \to 0$, this yields

$$\beta = \frac{(x_1 + x_2 + x_3 + x_4 + x_5)}{5} \tag{5.9}$$

Thus, we see that this filter is analogous to the FIR filter, unlike the median filters, and may be used in the same manner as the moving average is used in the gas turbine industry.

5.3 Numerical Simulations

The signals that are considered for preliminary numerical experiments are (1) step signal, (2) ramp signal, and (3) quadratic signal, as shown in Figure 5.1. Later in this chapter, more realistic gas path signals are considered. However, the numerical experiments conducted here allow selection of the value of the parameter K, which is critical to the performance of the myriad filter. The step signal simulates a single-fault event in the gas turbine. The ramp and quadratic signal simulate long-term deterioration. The signal is created using

$$x = x^0 + \alpha\varepsilon + \theta \tag{5.10}$$

where x^0 is the root signal, ε accounts for Gaussian noise, and θ for non-Gaussian outliers. Here α is a parameter called noise level that allows us to control the level of noise in the data. A Ψ filter performs the following operations that return the filtered signal from the corrupted signal:

$$y = \Psi(x) = \Psi(x^0 + \alpha\varepsilon + \theta) \tag{5.11}$$

The following error measure will be used to analyze the myriad filter performance over a sample of N points by comparing the filtered signal with the root signal:

$$MAE = \sum_{i=1}^{N} \frac{\left|(y - x^0)\right|}{N} \tag{5.12}$$

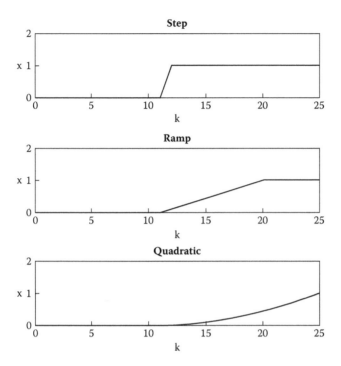

FIGURE 5.1
Test signals. (From Surender, V.P., and Ganguli, R., *Journal of Engineering for Gas Turbine and Power* 127(2):329–339, 2005. With permission.)

The mean absolute error (*MAE*) is widely used in signal processing and is particularly sensitive to the presence of outliers and edges in the data. The three ideal test signals are shown in Figure 5.1, which shows a signal from discrete time $k = 1$ until $k = 25$. These signals correspond to the step change in the measurements, a linear variation, and a quadratic variation. Five outliers are added to test the robustness of the filter to impulses: $\theta(7) = -1.0$, $\theta(18) = 1.0$, $\theta(10) = 0.75$, $\theta(14) = -0.75$, and $\theta(22) = -1.5$. Gaussian noise at various levels is added for the numerical experiments. Since the maximum value of the test signals is 1 and the minimum value is 0, by varying the noise level between 0 and 0.4, a wide variety of signal-to-noise ratios found in typical gas path measurements can be simulated as shown in Figure 5.2.

The effects of the linearity parameter K on *MAE* are on the step, ramp edge, and quadratic signals for the five-point filter. The simulation results for step, ramp, and quadratic signals for different values of K on *MAE* are shown in Figure 5.2 with various noise levels in the data. For the five-point filter, low values of K (such as $K = 0.01$) cause the myriad performance to become worse than the FIR at higher noise levels. For high values of K, such as $K = 5$,

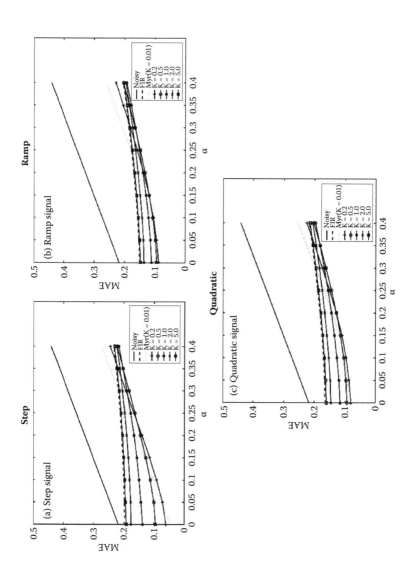

FIGURE 5.2

Effect of linearity parameter *K* on five-point myriad filter for *MAE*. (a) Step signal. (b) Ramp signal. (c) Quadratic signal. (From Surender, V.P., and Ganguli, R., *Journal of Engineering for Gas Turbine and Power* 127(2):329–339, 2005. With permission.)

the myriad filter approaches the FIR filter in performance. Since gas path signals have reasonably low levels of noise and have more possibility for outlier contamination, a value of $K = 0.1$ is selected.

5.4 Gas Turbine Transient Signal

The trajectories for a clean and deteriorated engine are presented during an acceleration process. The deteriorated engine had single-component faults in low-pressure turbine (LPT) and intermediate-pressure compressor (IPC) implanted, respectively. The signals of LPT fault and IPC fault on T2 and T6 during transients are obtained from [54]. These ideal signals are shown in Figures 5.3 and 5.4. The noise level α of 0.004 is the level of noise used in [54]. We use $\alpha = 0.004$ and $\alpha = 0.008$ to simulate conditions with normal data and low-quality data, respectively. Outliers are also added to the signals. The noisy signals are shown in Figures 5.5(a), 5.6(a), and 5.7(a). Tables 5.1 and 5.2 show the MAE values on the T2 measurements.

Results for both the clean and degraded engines are shown. Tables 5.3 and 5.4 show the MAE values for the T6 measurements. There is considerable reduction in the MAE using the myriad filter with $K = 0.1$, compared to the FIR filter. Again, results for both the clean and degraded engines are shown. The results for the weighted myriad filter shown in these tables will be discussed later in this chapter.

5.5 Weighted Myriad Algorithm

A problem in the myriad filter discussed above is in the selection of the linearity parameter K. Weighted myriad filters provide a way to avoid the selection of the linearity parameter.

Consider a set of observations $\{x_i\}_{i=1}^{N}$ and a set of filter weights $\{W_i\}_{i=1}^{N}$. Define the observation vector $x \underline{\Delta} [x_1, x_2, \ldots, x_N]^T$ and the weighted vector $w \underline{\Delta} [w_1, w_2, \ldots w_N]^T$.

For a given $K > 0$, the weighted myriad filter (WMyF) output is given by

$$\hat{\beta}_K(w, x) \underline{\Delta} \, myriad\,(K; w_1 \circ x_1, w_2 \circ x_2, \ldots, w_N \circ x_N) = \arg\min_\beta G_K(\beta, w, x) \quad (5.13)$$

The function

$$G_K(\beta, w, x) \, \underline{\Delta} \, \prod_{i=1}^{N} \left[K^2 + w_i (x_i - \beta)^2 \right] \quad (5.14)$$

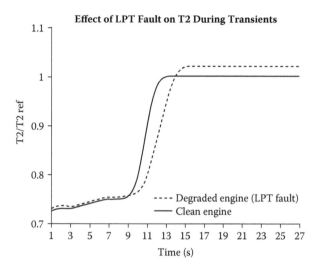

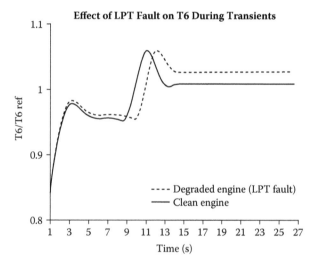

FIGURE 5.3
Clean and degraded signals during LPT fault on acceleration process. (From Surender, V.P., and Ganguli, R., *Journal of Engineering for Gas Turbine and Power* 127(2):329–339, 2005. With permission.)

is called the weighted myriad objective function, since it is minimized by the weighted myriad and $w_i \circ x_i$ denotes the weighting operation in Equation (5.13).

Using negative weights results in potential instability of the filter, so it is defined using positive weights: $w_i \geq 0$, $i = 1, 2, \dots N$. The WMyF output is the value of β at its global minimum of $G_K(\beta)$. For $w_i \geq 0$ and $K > 0$, $G(\beta) > 0$ for all β. Furthermore, $G_k(\beta)$ is a polynomial in β of degree $2N$. The myriad

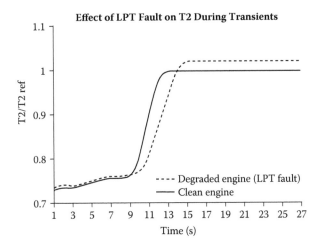

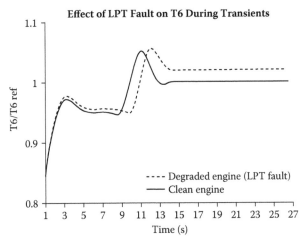

FIGURE 5.4
Clean and degraded signals during IPC fault on acceleration process. (From Surender, V.P., and Ganguli, R., *Journal of Engineering for Gas Turbine and Power* 127(2):329–339, 2005. With permission.)

$\hat{\beta}$ is defined as the minimum value of $G_K(\beta)$ and occurs at one of the roots of $G'_K(\beta)$.

$$G'_K(\beta) = 0 \tag{5.15}$$

Differentiating Equation (5.14) gives

$$G'_K(\beta) = \sum_{j=1}^{N} 2w_j \left(\beta - x_j\right) \prod_{\substack{l=1 \\ l \neq j}}^{N} \left[K^2 + w_l \left(\beta - x_l\right)^2 \right] \tag{5.16}$$

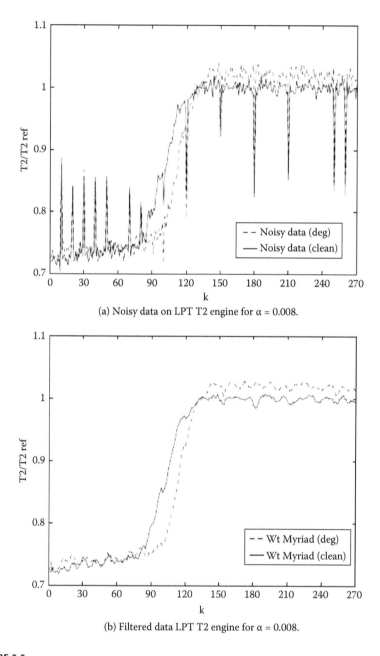

(a) Noisy data on LPT T2 engine for α = 0.008.

(b) Filtered data LPT T2 engine for α = 0.008.

FIGURE 5.5
Clean and degraded signals during LPT fault on acceleration process for *T2*. (a) Noisy data on LPT *T2* engine for α = 0.008. (b) Filtered data on LPT *T2* engine for α = 0.008. (From Surender, V.P., and Ganguli, R., *Journal of Engineering for Gas Turbine and Power* 127(2):329–339, 2005. With permission.)

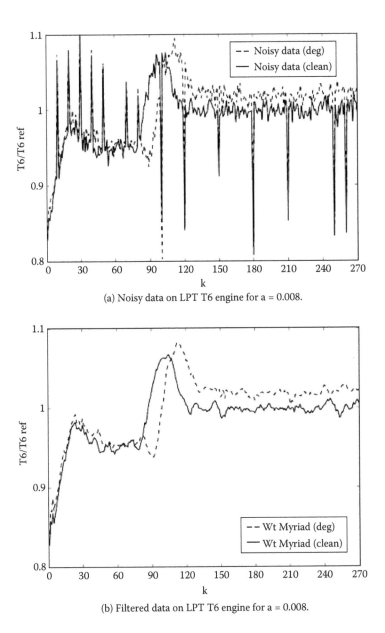

(a) Noisy data on LPT T6 engine for a = 0.008.

(b) Filtered data on LPT T6 engine for a = 0.008.

FIGURE 5.6
Clean and degraded signals during LPT fault on acceleration process for T6. (a) Noisy data on LPT *T6* engine for α = 0.008. (b) Filtered data on LPT *T6* engine for α = 0.008. (From Surender, V.P., and Ganguli, R., *Journal of Engineering for Gas Turbine and Power* 127(2):329–339, 2005. With permission.)

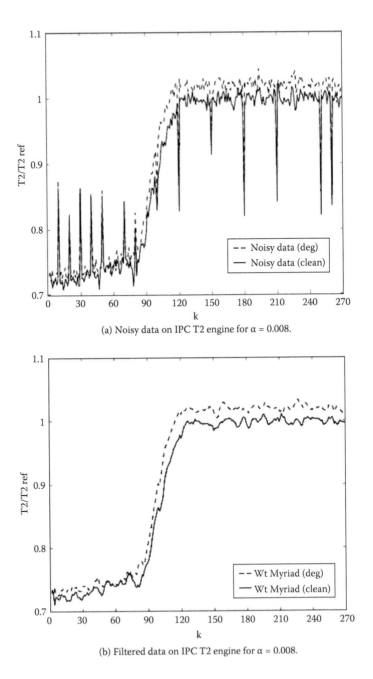

(a) Noisy data on IPC T2 engine for α = 0.008.

(b) Filtered data on IPC T2 engine for α = 0.008.

FIGURE 5.7
Clean and degraded signals during IPC fault on acceleration process for $T2$. (a) Noisy data on IPC $T2$ engine for $\alpha = 0.008$. (b) Filtered data on IPC $T2$ engine for $\alpha = 0.008$. (From Surender, V.P., and Ganguli, R., *Journal of Engineering for Gas Turbine and Power* 127(2):329–339, 2005. With permission.)

TABLE 5.1

MAE for *T2* Engine with $\alpha = 0.004$

	LPT *T2*		IPC *T2*	
Filter	Degraded	Clean	Degraded	Clean
FIR	0.1149	0.1101	0.1085	0.109
Myriad	0.074	0.0694	0.0666	0.069
WMyF	0.0549	0.0503	0.0483	0.0503

Source: From Surender, V.P., and Ganguli, R., *Journal of Engineering for Gas Turbine and Power* 127(2):329–339, 2005. With permission.

TABLE 5.2

MAE for *T2* Engine with $\alpha = 0.008$

	LPT *T2*		IPC *T2*	
Filter	Degraded	Clean	Degraded	Clean
FIR	0.1237	0.1136	0.1222	0.1277
Myriad	0.0829	0.0735	0.0817	0.0871
WMyF	0.0669	0.055	0.0635	0.0722

Source: From Surender, V.P., and Ganguli, R., *Journal of Engineering for Gas Turbine and Power* 127(2):329–339, 2005. With permission.

TABLE 5.3

MAE for *T6* Engine with $\alpha = 0.004$

	LPT *T6*		IPC *T6*	
Filter	Degraded	Clean	Degraded	Clean
FIR	0.1162	0.1111	0.1066	0.1148
Myriad	0.0681	0.0648	0.0588	0.0671
WMyF	0.0521	0.0498	0.0459	0.0519

Source: From Surender, V.P., and Ganguli, R., *Journal of Engineering for Gas Turbine and Power* 127(2):329–339, 2005. With permission.

TABLE 5.4

MAE for *T6* Engine with $\alpha = 0.008$

	LPT *T6*		IPC *T6*	
Filter	Degraded	Clean	Degraded	Clean
FIR	0.1260	0.1247	0.1203	0.1254
Myriad	0.0773	0.0791	0.0729	0.0782
WMyF	0.0634	0.0638	0.0593	0.0645

Source: From Surender, V.P., and Ganguli, R., *Journal of Engineering for Gas Turbine and Power* 127(2):329–339, 2005. With permission.

This is a polynomial of degree $2N - 1$ with as many as $2N - 1$ real roots. Using Equations (5.14) and (5.16), we get

$$G'_K(\beta) = 2G'_K(\beta) \sum_{j=1}^{N} \frac{w_j(\beta - x_j)}{K^2 + w_j(\beta - x_j)^2} \qquad (5.17)$$

Since $G'_K(\beta) > 0$, the filter output $\hat{\beta}$ must satisfy

$$\sum_{j=1}^{N} \frac{\dfrac{w_j}{K^2}(\hat{\beta} - x_j)}{1 + \dfrac{w_j}{K^2}(\hat{\beta} - x_j)^2} = 0 \qquad (5.18)$$

The above equation shows that even if the value of K is changed, the same filter output is obtained provided the weights are suitably scaled. Therefore,

$$\hat{\beta}_{K_1}(w_1, x) = \hat{\beta}_{K_2}(w_2, x) \qquad if \ \frac{w_1}{K_1^2} = \frac{w_2}{K_2^2} \qquad (5.19)$$

Thus, the filter output depends on w/K^2. Thus, the problem of finding the optimal linearity parameter K is avoided by using the weighted myriad filter. However, the problem of finding the appropriate value of K is now replaced by finding the appropriate weight vector. In general, the myriad is the global minima among several local minimum points of $G'(\beta)$. To compute the myriad, one needs to find the roots of polynomial $G(\beta)$, and test all the local minima to find the global minima.

5.6 Adaptive Weighted Myriad Filter Algorithm

For applications where the statistics of the signal are unknown or time varying, the adaptive algorithm is very useful. For gas path measurements, historical data are often used to obtain representative numbers for measurement uncertainty. However, it is likely that the uncertainty changes for different outliers and engines and for different operating conditions. The adaptive algorithm avoids the need for this simplification. An adaptive steepest descent method algorithm is used to optimize the filter weights. The update is computationally comparable to the least mean absolute deviation (LMAD) algorithm.

In order to find the optimal weights, the steepest descent method can be used to minimize the MAE cost function $J(w)$ [16], where

$$J(w, K) = E\{|y_K(w, x) - d|\} \qquad (5.20)$$

and $E\{.\}$ is the statistical expectation and d is the desired signal. Here d can be generated by an onboard model of the engine. The following algorithm is used to update the filter weights:

$$w_i(n+1) = P\left[w_i(n) - \mu \frac{\partial J}{\partial w_i}(n)\right] \qquad (5.21)$$

where $w_i(n)$ denotes the ith weight at the nth iteration, $\mu > 0$ is the step size of the update, and $P[.]$ is defined as

$$P[u] \triangleq u, \qquad \begin{cases} u, & u > 0 \\ 0, & u \le 0 \end{cases} \qquad (5.22)$$

In practice, $P[u]$ is set to a small positive value $u \le 0$. Differentiating Equation (5.20), we obtain

$$\frac{\partial J}{\partial w_i}(n) = E\left\{ \text{sgn}\left[y(n) - d(n)\right] \frac{\partial y}{\partial w_i}(n) \right\} \qquad (5.23)$$

Since lack of knowledge of signal statistics makes evaluation of statistical expectation E difficult, an instantaneous estimate for the gradient is used. Putting Equation (5.23) into Equation (5.21) yields

$$w_i(n+1) = P\left\{ w_i(n) - \mu \text{sgn}\left[e(n)\right] \frac{\partial y}{\partial w_i}(n) \right\} \qquad (5.24)$$

where $e(n) = y(n) - d(n)$ is the error at the nth iteration. To find $\frac{\partial y}{\partial w_i}$ for a given K, we can write Equation (5.17) as

$$G_K'(y) = 2G_K(y,w,x) \sum_{j=1}^{N} \frac{w_j(y - x_j)}{K^2 + w_j(y - x_j)^2} \qquad (5.25)$$

Consider K, the other weights, and input vector x are constant. Then

$$G'(y) = 2G(y,w_i) H(y,w_i) \qquad (5.26)$$

where

$$H(y,w_i) \triangleq \sum_{j=1}^{N} \frac{w_j(y - x_j)}{K^2 + w_j(y - x_j)^2} \qquad (5.27)$$

Since $G(y) = 0$,

$$G(y, w_i) H(y, w_i) = 0 \qquad (5.28)$$

Since $G(y, w_i) \geq 0$,

$$H(y, w_i) = 0 \tag{5.29}$$

Differentiating the above equation with respect to w_i yields

$$\frac{\partial H}{\partial y} \cdot \frac{\partial y}{\partial w_i} + \frac{\partial H}{\partial w_i} = 0 \tag{5.30}$$

$$\frac{\partial y}{\partial w_i} = -\frac{\dfrac{\partial H}{\partial w_i}}{\dfrac{\partial H}{\partial y}} \tag{5.31}$$

$$\frac{\partial H}{\partial w_i} = \frac{\partial}{\partial w_i} \sum_{j=1}^{N} \frac{w_j (y - x_j)}{K^2 + w_j (y - x_j)^2}$$

$$= \frac{1}{K^2} \left\{ \frac{(y - x_i)}{\left[1 + \dfrac{w_i}{K^2} (y - x_i)^2 \right]^2} \right\} \tag{5.32}$$

$$\frac{\partial H}{\partial y} = \frac{\partial}{\partial y} \sum_{j=1}^{N} \frac{w_j (y - x_j)}{K^2 + w_j (y - x_j)^2}$$

$$= \sum_{j=1}^{N} \frac{w_j}{K^2} \cdot \frac{1 - \dfrac{w_j}{K^2} (y - x_j)^2}{\left[1 + \dfrac{w_j}{K^2} (y - x_j)^2 \right]^2} \tag{5.33}$$

Putting Equation (5.32) and Equation (5.33) in Equation (5.31) we get

$$\frac{\partial y}{\partial w_i} = \frac{\left\{ \dfrac{-(y - x_i)}{\left[1 + \dfrac{w_i}{K^2} (y - x_i)^2 \right]^2} \right\}}{K^2 \left\{ \displaystyle\sum_{j=1}^{N} \dfrac{w_j}{K^2} \dfrac{1 - \dfrac{w_j}{K^2} (y - x_j)^2}{\left[1 + \dfrac{w_j}{K^2} (y - x_j)^2 \right]^2} \right\}} \tag{5.34}$$

Putting Equation (5.34) into Equation (5.24) yields

$$w_i(n+1) = P\left(w_i(n) + \mu \operatorname{sgn}\left[e(n)\right] \frac{\left\{\dfrac{(y-x_i)}{\left[1+\dfrac{w_i}{K^2}(y-x_i)^2\right]^2}\right\}(n)}{K^2\left\{\alpha + \displaystyle\sum_{j=1}^{N}\dfrac{w_j}{K^2}\dfrac{1-\dfrac{w_j}{K^2}(y-x_j)^2}{\left[1+\dfrac{w_j}{K^2}(y-x_j)^2\right]^2}\right\}(n)}\right)$$

(5.35)

We can remove the denominator from the update term in the above equation, as it does not affect the direction of the gradient estimate or the values of the final weights. This leads to

$$w_i(n+1) = P\left(w_i(n) + \mu \operatorname{sgn}\left[e(n)\right]\left\{\frac{(y-x_i)}{\left[1+\dfrac{w_i}{K^2}(y-x_i)^2\right]^2}\right\}(n)\right)$$

(5.36)

where $e(n) = y(n) - d(n)$ is the error at the nth iteration, d is the desired signal, $y(n)$ is the WMyF output at the nth iteration, n is the number of iterations, μ is the step size, and $w_i(n)$ denotes the ith weight at the nth iteration. $P(u)$ is set to a small positive value ε if $u < 0$. The derivation of the algorithm is condensed from [16]. The above algorithm is started with weights equal to 1 at iteration $n = 1$, $K = 0.1$, and $\mu = 5.0 \times 10^{-3}$. The proposed schematic of the adaptive filter is shown in Figure 5.8.

In applications where the non-Gaussian process naturally arises in practice, the myriad-based algorithms lead to significant advantages. In gas turbine

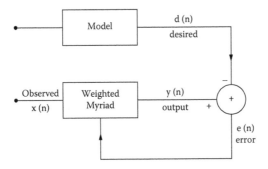

FIGURE 5.8
Schematic for adaptive filter optimization. (From Surender, V.P., and Ganguli, R., *Journal of Engineering for Gas Turbine and Power* 127(2):329–339, 2005. With permission.)

applications, the presence of outliers in the data is well known. The weighted myriad filters have the potential to perform significantly better than both linear and median filters in Gaussian and non-Gaussian environments. For applications where the statistical characteristics of the underlying signals may be unknown or varying, adaptive algorithms are used. A disadvantage of the myriad filter is its complicated definition and implementation compared to the linear filters. However, once implemented, this is hidden from the user.

5.7 Numerical Simulations

This adaptive algorithm was evaluated through a computer simulation example involving denoising gas turbine measurements contaminated with Gaussian noise. The trajectories for a clean and deteriorated engine are considered for this problem. In our simulation example, the degraded signal and clean signal were chosen that had single-component faults: low-pressure turbine (LPT) and intermediate-pressure compressor (IPC) faults. The signal is corrupted by adding noise level $\alpha = 0.004$ [54], yielding the noisy observed signal. A further noise level of $\alpha = 0.008$ was also used.

The objective of this adaptive algorithm is to train the weighted myriad to converge the filter weights that minimize the absolute value of the error signal $e(n)$ between the filter output signal $y(n)$ and desired signal $d(n)$. This adaptive algorithm is a simple, fast, and practical weighted myriad filter algorithm. This achieves a lower *MAE* than the FIR and sample myriads at comparable convergence speeds, as shown in the results in Tables 5.1–5.4.

The weighted myriad filter has the best outputs and lowest mean errors. The effects of the five-point filter are plotted with $\alpha = 0.008$ for the noisy data and weighted myriad in Figures 5.5–5.7. In this case, we took both the degraded signal and the clean signal for a *T2* LPT engine sensor signal. Figure 5.5(a) shows the noisy data, for both degraded and clean engines. The weighted myriad filtered data are shown in Figure 5.5(b) for the degraded and clean engine signal.

Similarly, the effects of LPT *T6* for both degraded and clean signals are plotted along with noisy data in Figure 5.6(a) and for the weighted myriad data in Figure 5.6(b). The results for the IPC engine sensor signal for the *T2* engine for degraded and clean signals are also plotted in Figure 5.7. Figure 5.7(a) shows the noisy data, for both degraded and clean engines, and Figure 5.7(b) shows the filtered data. From these figures, and results in Tables 5.1–5.4, it is clear that considerable noise reduction is obtained using the weighted myriad filter. The outcomes of different filters, such as FIR, sample myriad, and weighted myriad, are compared and studied.

From Figure 5.9(a), we can see that the weighted myriad filter shows much better performance and outlier removal than the FIR filter,

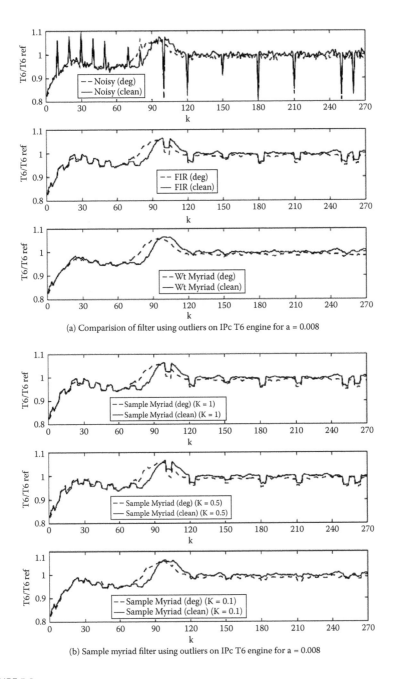

(a) Comparision of filter using outliers on IPc T6 engine for a = 0.008

(b) Sample myriad filter using outliers on IPc T6 engine for a = 0.008

FIGURE 5.9
Comparison of FIR and myriad filters. (a) Comparison of filters using outliers on IPC *T6* engine for α = 0.008. (b) Sample myriad filter using outliers on IPC *T6* engine for α = 0.008. (From Surender, V.P., and Ganguli, R., *Journal of Engineering for Gas Turbine and Power* 127(2):329–339, 2005. With permission.)

which is not able to remove the outliers. The sample myriad filter results are plotted in Figure 5.9(b) for different values of K. It is clear from the results that the sample myriad can be controlled by adjusting the linear parameter K, whereas the adaptive weighted myriad filter provides a way to avoid the selection of the linear parameter K. However, a low value of $K = 0.1$ appears to work quite well for the signals in this study. Note that the sample myriad used the value $K = 0.1$, and the values of K that give good results may change for different types of signal and noise characteristics.

This chapter therefore offers two approaches for denoising gas path signals. The first is based on the sample myriad and uses a suitable value of K based on representative signals and known statistics of the system. Our results show that $K = 0.1$ works quite well for signals polluted by outliers. This approach is suitable for ground-based diagnostic systems. The sample myriad filter is implemented as a MATLAB® program using built-in functions that calculate the roots of polynomials and are simple and fast. The second approach is the powerful weighted myriad approach that adapts to changes in both signals and their statistics, and is especially suitable for online applications. The adaptive myriad is implemented using the weight update Equation (5.36) using MATLAB. While the adaptive myriad is a useful tool for outlier-contaminated gas path measurements, the optimum weightings are not known until all the available data are processed. Thus, there is a lag caused by the filter. Engine faults result in a change in engine state measured over many samples and should not affect the myriad filter. Abrupt faults will not be impacted by filters with low window length, and gradual faults that develop slowly over time will also not be impacted. However, any intermittent faults may be impacted, as they can appear as outliers to the filter.

5.8 Summary

In this chapter, we looked at the myriad filter as a substitute for the moving average filter, which is widely used in the gas turbine industry. Results with transient gas path measurement signals show that the myriad filter has good noise reduction qualities when compared to moving average FIR filters. There is considerably more noise reduction obtained by the myriad filter than with the FIR filters. The results show that there is good outlier removal by the weighted myriad filter, which is an important element of a filter. The adaptive weighted myriad filter shows even greater noise reduction than the sample myriad and FIR filter in the gas turbine diagnosis.

The use of such a filter prior to performing fault isolation and detection functions is likely to lead to improved performance of the diagnostic system. This filter is particularly suited for online and real-time gas turbine diagnostic systems where engine data, both steady state and transient, are continuously and rapidly available.

6

Trend Shift Detection

The last few chapters focused on noise removal from gas turbine signals. The signal processing algorithms ranged from simple "gentle smoothers" like the CWIM filter suitable for trend plots to the sophisticated adaptive myriad filter for real-time transient data analysis. In this chapter, we address the problem of locating the discrete time or epoch when the sudden change in the gas turbine signal occurs. Trend shift detection is posed as a two-part problem: filtering of the gas turbine measurement deltas followed by the use of edge detection algorithms. Measurement deltas are deviations in engine gas path measurements from a good baseline engine and are a key health signal used for gas turbine performance diagnostics. Just like in the previous chapters, the measurements used in this chapter are exhaust gas temperature, low rotor speed, high rotor speed, and fuel flow, which are called cockpit measurements and are typically found on most commercial gas turbines. In this chapter, a cascaded recursive median (RM) filter, of increasing order, is used for the purpose of noise reduction and outlier removal, and a hybrid edge detector that uses both gradient and Laplacian of the cascaded (RM) filtered signal is used for the detection of step change in the measurements. Simulated results with test signals indicate that cascaded RM filters can give a noise reduction of more than 38% while preserving the essential features of the signal. The cascaded RM filter also shows excellent robustness in dealing with outliers, which are quite often found in gas turbine data, and can cause spurious trend detections. Suitable thresholding of the gradient edge detector coupled with the use of the Laplacian edge detector for cross-checking can reduce the system false alarms and missed detection rate. Further reduction in the trend shift detection false alarm and missed detection rate can be achieved by selecting gas path measurements with higher signal-to-noise ratios.

The FIR-median hybrid filter discussed in Chapter 4 worked well only when the length of the window was quite large and had a 10-point delay in processing a newly arriving point. Such a system would be useful if the data were received at over 10 points per flight by the monitoring system. In general, more than two comparable readings in a single aircraft flight is very unusual. Timely alerting would require a near-real-time mode of transmission that can be very expensive.

In this chapter, we use the cascaded recursive median filter to remove noise from signals. Unlike the nonrecursive filters discussed in the earlier chapters, the recursive filters converge very fast and do not need repeated passes. They are also good at removing outliers like other median filters.

The use of recursive median filters in cascades is a recent research development in the field of nonlinear signal processing. In addition, we use a gradient and Laplacian-based edge detector to minimize false alarms and missed detections. This chapter is based on the concept of coupling recursive median filters with an edge detector, which was proposed for gas turbine diagnostics by Ganguli and Dan [55].

6.1 Problem Formulation

Consider the four basic (cockpit) measurements from an aircraft gas turbine engine: *EGT, WF, N1,* and *N2.* Almost all engines, including older engines, have these measurements. Some newer engines may be fitted with additional pressure and temperature sensors. Therefore, we concentrate on these four measurements. Deltas for each of these are obtained by subtracting the baseline measurements for a good engine from the actual measurement. The baseline measurements often come from an engine model, and various correction factors are used to reduce the measured data to standard sea level conditions. But these gas path measurement deltas contain high levels of uncertainty due to sensor errors, high-frequency noise, and modeling assumptions.

For a typical engine, the measurement deltas slowly increase with time due to deterioration as the number of flights increases. While deterioration accumulates over many flights, faults are more abrupt or step changes. For a trend shift to be detectable, it must show up beyond the scatter band caused by noise for the measurements. In general, the trend shift can be any amount depending on the impending fault. In this chapter, a step change in measurement deltas of 2% or more is interpreted as a large enough trend shift implying a single-fault event. Thus, we can write the measurement deltas as follows:

$$z = z^0 + \theta \tag{6.1}$$

where θ is noise and z^0 is the baseline measurement delta. Hence, given the real noisy measurement deltas, our problem is *fast detection of trend shift in the presence of noise in the data.* Noise is added to the simulated measurement deltas using the typical standard deviations for ΔEGT, ΔWF, $\Delta N1$, and $\Delta N2$ as 4.23°C, 0.50%, 0.25%, and 0.17%, respectively. These values are representative of airline data and are obtained from Lu et al. [11, 31]. It is also possible that there are non-Gaussian outliers or wild points, as stated by Lu et al., in the data. We have used these noise statistics in previous chapters.

Note that the problem formulation above is idealized and does not account for engine-to-engine variation, measurement bias, and gradual engine deterioration, among others. However, the problem allows for an illustration of the proposed algorithms.

6.2 Image Processing Concepts

Our objective is to detect trend shifts in gas turbine measurement deltas. To do so, we borrow ideas from image processing [56]. Images are two-dimensional signals composed of pixels with different levels of brightness. Images are often contaminated with Gaussian as well as impulsive noise. The need for detection of edges (pixels with higher brightness than the neighborhood) has led to significant research on nonlinear filtering and edge detection.

Typical edge detection methods [57–59] use gradient (first derivative) or Laplacian (second derivative) operators to detect edges in images that are essentially discontinuities in the pixel values. In many machine vision applications, it is useful to separate out the regions of the image corresponding to the objects of interest, from the regions of the image corresponding to the background. At the edge, the magnitude of the gradient peaks and the Laplacian goes through a zero crossing. Thresholding the gradient is a way to perform the segmentation of the image on the basis of different regions or colors in the foreground and background of the image. Thus, a grayscale image can be interpreted in black and white.

However, in order to use such derivative operators for edge detection, the image needs to be preprocessed using some smoothing filter to suppress noise. This is because noise in the image gets amplified due to derivative operations. Unfortunately, linear smoothing methods blur the sharp edges in the image, and are not good at removing impulsive noise. Therefore, nonlinear filters such as median filters are often used to preprocess images [60–70].

This image processing research is applicable to the diagnostics problem because the gas turbine measurement delta signal can be viewed as a one-dimensional image. In the following sections, we introduce the concepts of the cascaded recursive median filters and edge detection in greater detail.

6.3 Median Filter

Recall that median filters are an important class of nonlinear filters. Nonlinear filters are multiscale in nature and possess the special ability of reducing noise without affecting the various features of the signal, which may represent a fault in the engine. An N-point median filter takes N points surrounding the central point and gives their median as the output; i.e., if z_k denotes the input signal, then the output of the median filter is

$$y_k = median(z_{k-n}, z_{k-n+1}, \dots, z_k, \dots, z_{k+n-1}, z_{k+n}) \tag{6.2}$$

where $N=2n+1$ is the window length of the filter. Since the median does not cause much blurring to edges, it can be applied iteratively. However, a very large number of iterations can be required by the median filter to converge to a root signal [60]. A root signal is a signal that does not change on further passes of the median filter, which means that for the entire signal

$$z_k = median(z_{k-n}, z_{k-n+1}, \ldots, z_k, \ldots, z_{k+n-1}, z_{k+n}) \qquad (6.3)$$

In addition, individual spikes do not affect the median value, so median filters remove impulsive noise quite well. For example, the median filter can discard gross outliers. Since a median filter takes previous as well as future input values for calculating a particular output, it has an associated time lag. But at the same time, it is much more effective than a linear filter in eliminating high-frequency Gaussian noise while preserving the essential signal features.

6.4 Recursive Median Filter

A modified version of these median filters is the recursive median (RM) filter [61–63]. Recursive median filters possess superior noise attenuation capability compared to their nonrecursive counterparts [62]. An RM filter uses some previous output values, instead of previous values for arriving at the next output, i.e., for an RM filter,

$$y_k = median(y_{k-n}, y_{k-n+1}, \ldots, x_k, \ldots, z_{k+n-1}, z_{k+n}) \qquad (6.4)$$

where $N=2n+1$ is the window length of the filter. Some fundamental properties of RM filters are:

1. Any input signal reduces to a root signal after one or very few RM filter steps, or equivalently, the output will not be modified by further application of the same filter. The RM filter is therefore more efficient than the median filter that can require many iterative passes to converge to the root. Recursive filters also ease hardware implementation.

2. The RM filter output is always made up of monotone sequence (edges) linked together with "constant neighborhoods" having a length of at least $n + 1$ if the window length of the filter is $2n + 1$.

Recursive median filters also have a higher immunity to impulsive noise or outliers in the data than median filters. Recursive median filters can be improved further when they are arranged in a cascade of increasing order.

6.5 Cascaded Recursive Median Filter

Bangham [64] observed that better noise removal is possible, compared to the recursive median filter, if we apply more than one RM filter sequentially on the same set of data points with the window lengths of the filter increasing. These are called cascaded RM filters and also the recursive median sieve [64–69]. Cascaded recursive median filters also make for easier software [65] and hardware implementation [66]. Alliney [67] gives analytical results showing the advantages of the cascaded RM filters. Consider one m-point RM filter and one n-point RM filter cascaded together. Passing the input signal through the m-point RM filter and the corresponding output through the n-point RM filter gives the final output signal, with quite large noise reduction and at the same time maintaining good feature representation. The results obtained are best if we use RM filters of increasing window size, i.e., $2n+1$, $n=1, 2, \ldots$, arranged in a cascade. For example, a three-point and five-point recursive median filter can be cascaded as follows:

$$y_k^1 = median\left(y_{k-1}^1, z_k, z_{k+1}\right)$$

$$y_{k+1}^1 = median\left(y_k^1, z_{k+1}, z_{k+2}\right)$$

$$y_{k+2}^1 = median\left(y_{k+1}^1, z_{k+2}, z_{k+3}\right)$$

(6.5)

$$y_k^2 = median\left(y_{k-2}^2, y_{k-1}^2, y_k^1, y_{k+1}^1, y_{k+2}^1\right)$$

The use of median filters of increasing order, arranged in a cascade, is especially profitable when one wants to remove disturbances superimposed on rectangular impulse trains or sharp changing signals. However, the use of the five-point RM filter results in a three-point time delay. This can be observed by considering the following logic:

1. The five-point RM output y_k^2 is a function of y_{k+2}^1.
2. The output y_{k+2}^1 from the three-point RM filter is a function of z_{k+3}.

Recently, Yli-Harja et al. [70] have shown that cascaded median filters can be implemented in hardware in a straightforward and compact manner. They are therefore useful for online applications also. Alliney [67] also mentions that the resulting signals after passing through the cascaded recursive median filters appear to be very satisfactory from a visual point of view. He speculates that some relationship could exist between the nonlinear filtering algorithms and the human visual cognition process. The human visual system tends to give very high importance to edges in images and signals, in a manner similar to that of the nonlinear filters.

6.6 Edge Detection

Edge detectors are a collection of image preprocessing methods used to locate changes in the image intensity function. Images are composed of pixels and edges are pixels where this function (brightness) changes abruptly. There are many methods for edge detection in images. However, among them the most widely used are the gradient-based edge detector and the Laplacian edge detector. These methods allow us to locate changes in the intensity function using derivatives. The gradient and Laplacian edge detectors are defined below.

6.6.1 Gradient Edge Detector

An edge is a monotone sequence surrounded by constant neighborhoods of different values. As a result, a sharp peak in the gradient characterizes an edge. In general, comparing the magnitude of the gradient to a threshold can identify candidate edge points in a signal. Thresholding ensures that all points having a local gradient above the threshold must represent an edge. Thresholds are typically set based on an estimated signal-to-noise ratio. The Canny edge detector [57], for example, uses gradients for edge detection. If the threshold is set low, then all edge points in a signal will be detected. However, nonedge points, including regions of high noise, will also be falsely detected. These false alarms can be minimized by using a filtering operation that removes noise but leaves the edge intact.

6.6.2 Laplacian Edge Detector

The gradient at an edge reaches a maximum. Similarly, the Laplacian at an edge equals zero. Therefore, there is a change in the sign of the Laplacian before and after the edge occurs. In general, it is much easier and more precise to find a zero crossing than a maximum point. In addition, while the gradient depends on the steepness of the edge, the Laplacian does not. The Marr edge detector [58] uses the zero crossing of the Laplacian for edge detection.

A key problem in edge detection is that the gradient and Laplacian tend to amplify the effect of high-frequency noise in the data. In addition, the presence of impulsive noise or outliers can cause spurious edges to be detected.

Therefore, a smoothing or filtering method is generally used on the signal before performing edge detection. We will use the cascaded recursive median filter for smoothing the gas path measurement deltas. We also use the gradient and Laplacian edge detector simultaneously for improved edge detection accuracy. Chou and Bennamoun [59] recently suggested the use of such a hybrid edge detector combining the first- and second-order differential edge detectors. They showed that for two-dimensional medical images their combination of two differential edge detectors gave accurate edge localization and showed robustness to noise.

6.7 Numerical Simulations

Simulated data for the test signal are used to test the cascaded RM filter for noise reduction. The filtered test signal is then used to test the combination of the gradient edge detector and the Laplacian edge detector for trend shift detection.

6.7.1 Test Signal

The test signal stretches over 20 discrete time points, and it assumes that some fault arises in the engine at discrete time $k=11$ and continues until $k = 20$, resulting in an individual 2% change in the ΔWF, $\Delta N1$, and $\Delta N2$ measurements and a 13.6°C change in ΔEGT. The ΔEGT signal is selected in terms of actual temperature, as percent changes are unavailable from the literature. A 13.6°C change in ΔEGT corresponds to a 2% high-pressure compressor (HPC) performance loss using faults defined in Lu et al. [11], Volponi et al. [12], and Ganguli [13]. The ideal test signals, along with a signal with added Gaussian noise, are shown in Figures 6.1–6.4. The standard deviations for the Gaussian noise added to ΔWF, $\Delta N1$, $\Delta N2$, and ΔEGT are 0.5%, 0.25%, 0.17%,

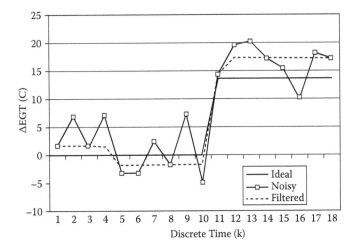

FIGURE 6.1
Ideal, noisy, and cascaded recursive median filtered signal for ΔEGT. (From Ganguli, R., and Dan, B., *Journal of Engineering for Gas Turbine and Power* 126(1):55–61, 2004. With permission.)

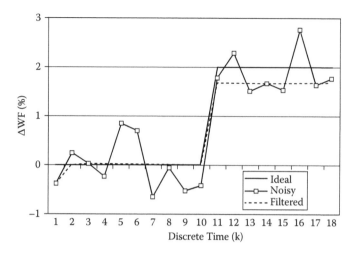

FIGURE 6.2
Ideal, noisy, and cascaded recursive median filtered signal for ΔWF. (From Ganguli, R., and Dan, B., *Journal of Engineering for Gas Turbine and Power* 126(1):55–61, 2004. With permission.)

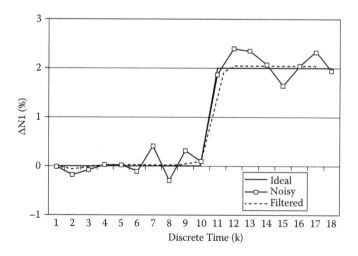

FIGURE 6.3
Ideal, noisy, and cascaded recursive median filtered signal for ΔN1. (From Ganguli, R., and Dan, B., *Journal of Engineering for Gas Turbine and Power* 126(1):55–61, 2004. With permission.)

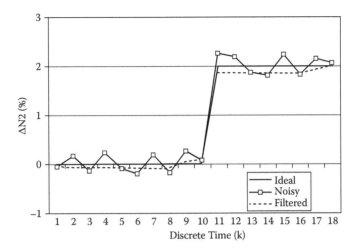

FIGURE 6.4
Ideal, noisy, and cascaded recursive median filtered signal for ΔN2. (From Ganguli, R., and Dan, B., *Journal of Engineering for Gas Turbine and Power* 126(1):55–61, 2004. With permission.)

and 4.23°C, respectively. These values are obtained from a comprehensive study of airline data and are reported by Lu et al. and Ganguli.

6.7.2 Noise Reduction

Figures 6.1–6.4 show the results of passing the noisy data through the three-point RM filter and the five-point RM filter, cascaded together. It is clear that the cascaded RM filter, of increasing order, removes a considerable amount of high-frequency noise, while preserving the sharp trend shifts. Since a trend shift can identify the temporal location of a fault, preserving them is very important for fault detection. In addition, by removing noise while preserving signal features, the cascaded RM filter enables better visualization of the signal. Figures 6.1–6.4 give a qualitative idea of noise reduction in the data due to passage through the cascaded recursive median filter.

To obtain a quantitative idea of the noise reduction, we look at the root mean square (RMS) error in the signal, which is a measure of the difference between the noisy or filtered and the ideal signal and is given as

$$\Theta = \frac{1}{M}\sqrt{\sum_{i=1}^{M}\left(z_i - z_i^0\right)^2} \qquad (6.6)$$

The error is a measure of noise in the signal, M is the number of points in the signal sample, and z_i is the ith measurement delta. If Θ is zero, all the noise has been eliminated, and the real signal is identical to the ideal signal.

To obtain the error measure Θ, we create 5000 samples of noisy data for the test signal. The error is then calculated as the average value for all the noisy signals. It is found that for all four measurements there is a reduction in the noise Θ of about 38% after filtering, compared to the noisy signal. It should be noted that this substantial noise reduction is obtained while preserving the trend shifts in the signal, which are needed for fault detection.

6.7.3 Outlier Removal

While all other results in this chapter are obtained assuming Gaussian noise, we illustrate the power of the cascaded RM filter to remove outliers in Figure 6.5. An outlier can be defined as a data point that appears to be inconsistent with the rest of the data. Here the noisy signal for ΔWF in Figure 6.2 is further contaminated by adding the value 1% at $k=5$ and subtracting 1% at $k = 15$ to the noisy signal. Since the standard deviation for ΔWF is 0.5, these are equivalent to a 2σ addition and subtraction. It is clear from Figure 6.5 that the cascaded RM filter discards the outlier points easily.

The results of using the exponential average filter are also shown in Figure 6.5. This filter has the form

$$y_k = az_k + (1-a)y_{k-1} \tag{6.7}$$

The results in Figure 6.5 use $a=0.25$. It can be observed that the exponential average creates a smoothing out of the trend shift. The cascaded RM filter has an advantage in removing outliers and maintaining the trend feature

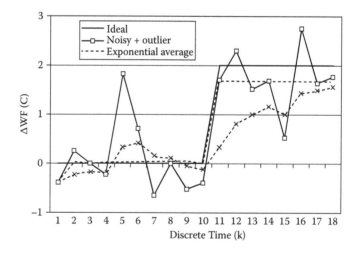

FIGURE 6.5
Noisy signal with outliers at $k=5$ and $k=15$ along with cascaded recursive median filtered signal and exponential average filtered signal. (From Ganguli, R., and Dan, B., *Journal of Engineering for Gas Turbine and Power* 126(1):55–61, 2004. With permission.)

and temporal information in the filtered data. Note, however, that Figure 6.5 does not indicate the time delay of three points in the RM filter when compared to the exponential filter.

6.8 Trend Shift Detection

A key factor in any edge detection algorithm is establishing where the detection threshold should be set. The measurement deltas entering the edge detector are preprocessed by the cascaded RM filter. For all results, we use a combined gradient/Laplacian edge detector. This edge detector detects an edge if the gradient exceeds a predetermined threshold value and the Laplacian changes sign at that point. Figure 6.6 shows a schematic representation of the trend shift detection algorithm.

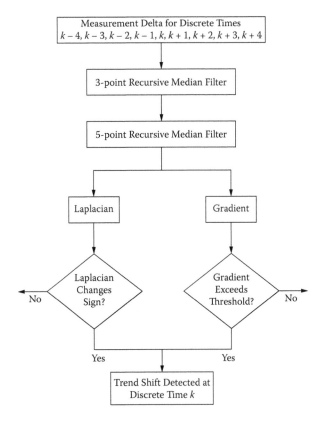

FIGURE 6.6
A schematic view of trend shift detection algorithm. (From Ganguli, R., and Dan, B., *Journal of Engineering for Gas Turbine and Power* 126(1):55–61, 2004. With permission.)

To understand the mechanism of edge detection, consider the gradient and Laplacian of the ΔWF measurement (Figure 6.2) shown in Figures 6.7 and 6.8, respectively. Note that the trend shift takes place between discrete time $k-1$ and k of 10 and 11, respectively.

For discrete signals, we need to calculate the derivative by finite difference approximation. Using a backward difference calculation of the gradient, we have

$$\nabla_k = z_k - z_{k-1} \tag{6.8}$$

where z_k and z_{k-1} are the measurement deltas at discrete time k and $k-1$. Therefore, the gradient shows a peak at $k = 11$ for the trend shift. The Laplacian is then calculated as

$$\nabla_k^2 = \nabla_k - \nabla_{k-1} \tag{6.9}$$

The Laplacian changes sign between k and $k+1$ of 11 and 12, respectively. Therefore, we can say that a trend shift occurs at discrete time k if

$$|\nabla_k| > T \quad \text{and} \quad sign\left(\nabla_k^2\right) \neq sign\left(\nabla_{k+1}^2\right) \tag{6.10}$$

where T is a threshold selected from numerical experiments discussed later in this chapter. There is an additional time delay of one point in our formulation above since ∇_{k+1}^2 is required.

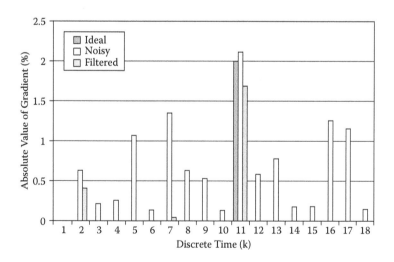

FIGURE 6.7
Absolute value of gradient for ideal, noisy, and cascaded recursive median filtered signal in Figure 6.3. (From Ganguli, R., and Dan, B., *Journal of Engineering for Gas Turbine and Power* 126(1):55–61, 2004. With permission.)

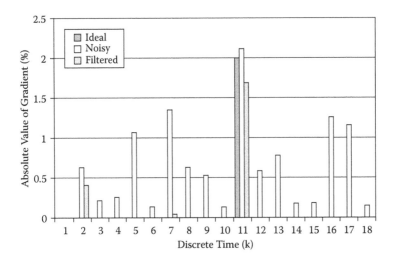

FIGURE 6.8
Laplacian of ideal, noisy, and cascaded recursive median filtered signal in Figure 6.3. (From Ganguli, R., and Dan, B., *Journal of Engineering for Gas Turbine and Power* 126(1):55–61, 2004. With permission.)

The use of the RM filters introduces a three-point time delay, as discussed earlier. Thus, the total time delay for the edge detection algorithm is four points.

The ideal data in Figure 6.7 show a trend shift at point 11, which is validated at point 12 by the zero crossing of the Laplacian in Figure 6.8. However, the noisy data show gradient values that are nonzero at many points, and show several zero crossings of the Laplacian. For the filtered signal, the noise is greatly reduced, and so are the spurious edge indicators for the gradient and the Laplacian. All future results are obtained with the filtered signal and the algorithm defined in Figure 6.6.

6.8.1 Threshold Selection

To empirically obtain a threshold on the gradient edge detector, we take 5000 samples of noisy data for each measurement, assume a threshold T, and calculate the number of false alarms and the missed detections. A missed detection occurs when the trend shift, which has occurred, is not detected. A false alarm occurs when a point where the trend shift does not occur is flagged as a trend shift point. The thresholds are varied over a range of values to evaluate the number of false alarms and missed detections given by the edge detector. The false alarms are measured by removing the trend shift in the test signal, which makes it a signal for a healthy engine with added noise.

Figures 6.7–6.10 show the results of the numerical experiments for ΔEGT, ΔWF, $\Delta N1$, and $\Delta N2$, respectively. As the threshold value is increased,

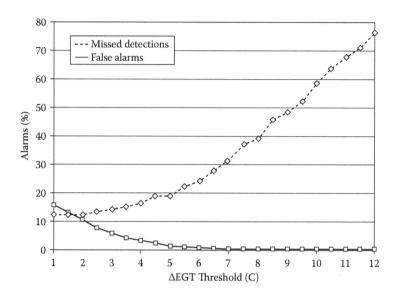

FIGURE 6.9
Missed detections and false alarms for trend shift detection with varying values of threshold on ΔEGT. (From Ganguli, R., and Dan, B., *Journal of Engineering for Gas Turbine and Power* 126(1):55–61, 2004. With permission.)

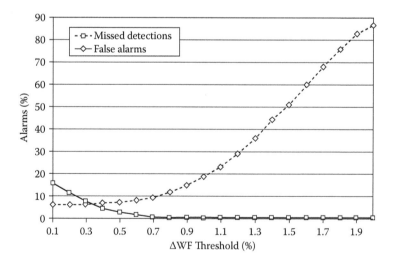

FIGURE 6.10
Missed detections and false alarms for trend shift detection with varying values of threshold on ΔWF. (From Ganguli, R., and Dan, B., *Journal of Engineering for Gas Turbine and Power* 126(1):55–61, 2004. With permission.)

the number of false alarms decreases but the number of missed detections increases. Measurements with better signal-to-noise ratio (SNR) can be used to obtain threshold values, which gives both reduced false alarms and missed detection rates. In our study, the SNRs for ΔEGT, ΔWF, $\Delta N1$, and $\Delta N2$ are 3.215, 4, 8, and 11.764, respectively. These nondimensional values are obtained by dividing the maximum trend shift magnitude by the standard deviation of a given measurement (for ΔEGT, 13.6/4.23; for ΔWF, 2/0.50; for $\Delta N1$, 2/0.25; and for $\Delta N2$, 2.0/0.17).

A quantitative idea of these numerical experiments can be obtained from Table 6.1. Analyzing Figures 6.9–6.12 and the underlying data, a suitable threshold value is chosen for each case, such that there are no false alarms

TABLE 6.1

False and Missed Detections with Thresholds Obtained from Numerical Experiments

Measurements	Threshold	Missed Detections (%)	False Alarms (%)
EGT	11.5°(C)	70.72	0.00
WF	1.40%	43.82	0.00
N1	0.60%	0.00	0.00
N2	0.40%	0.00	0.00

Source: Ganguli, R., and Dan, B., *Journal of Engineering for Gas Turbine and Power* 126(1):55–61, 2004. With permission.

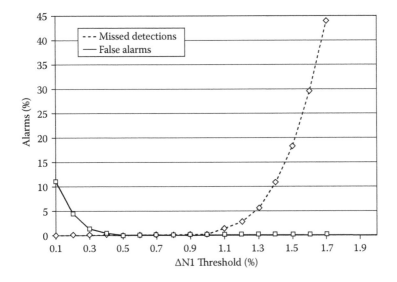

FIGURE 6.11
Missed detections and false alarms for trend shift detection with varying values of threshold on $\Delta N1$. (From Ganguli, R., and Dan, B., *Journal of Engineering for Gas Turbine and Power* 126(1):55–61, 2004. With permission.)

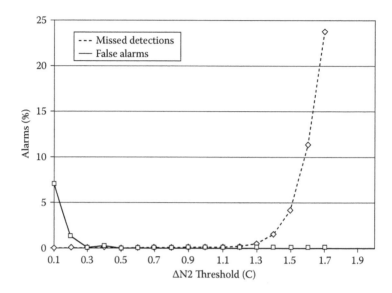

FIGURE 6.12
Missed detections and false alarms for trend shift detection with varying values of threshold on $\Delta N2$. (From Ganguli, R., and Dan, B., *Journal of Engineering for Gas Turbine and Power* 126(1):55–61, 2004. With permission.)

and minimum missed detections for the 5000 data samples used in these evaluations.

6.8.2 Testing of Trend Detection Algorithm

Once the threshold values have been selected, the trend detection algorithm defined in Figure 6.6 can be tested. Using the threshold values in Table 6.1, the percentage of false alarms and missed detections is calculated over 5000 new samples of noisy data and are given in the Table 6.1. These noisy samples are different from those used for creating the thresholds. For the ΔEGT and ΔWF measurement, there are a considerable number of missed detections, if the threshold is selected to minimize false alarms. However, for the rotor speeds $\Delta N1$ and $\Delta N2$, we can obtain thresholds such that the false and missed detections are both zero. The high rotor speed $\Delta N2$ turns out to be the best measurement to monitor for trend detection.

It should be noted that other measurements besides rotor speed might provide an indication of engine distress for certain type of faults. In general, rotor speed shifts are small, and so the true signal-to-noise ratio would be different for different faults. The ΔEGT and ΔWF signals are then more likely to reflect engine health. In general, the establishment of detection thresholds is a trade-off between false alarms and missed detections. These thresholds can also be set or tuned by the end user or by some automated system based on historical data. Individual sensor signal-to-noise ratios will determine the achievable false alarm and missed detection ratio.

Finally, we point out that one of the benefits of the trend shift detection approach proposed in this chapter is that it will work regardless of whether the trend shift is from zero or from some other nonzero quasi-steady-state measurement value. This is because of the use of gradient information for edge detection.

It should also be pointed out that there are specific trend shift cases that the algorithm developed in this chapter alone would not detect. This includes intermittent shifts that the median filter would discard as outliers. It also includes gradual trend shifts that occur over several samples that the edge detector would not necessarily detect. The cascaded RM filter with edge detection is best suited for sharp trend shifts in gas path measurements.

6.9 Summary

Fast and effective trend shift detection requires filtering of the data for removing the high-frequency noise while preserving the sharp edges. Among nonlinear filters, cascaded recursive median filters, of increasing order, are found to have the capability of noise removal with accurate signal feature preservation. They are also very fast converging; i.e., only one pass is required to obtain the accurate root signal, and are well suited for software and hardware implementations. For testing the filters, simulated faulty data, indicating the effect of the fault in the engine on ΔEGT, ΔWF, $\Delta N1$, and $\Delta N2$, are passed sequentially through a three-point RM filter and five-point RM filter. A substantial noise reduction of about 38% is found after a single pass. The task of trend shift detection is accomplished by using a combination of a gradient edge detector and a Laplacian edge detector. This combination is also found to be very effective. A suitable choice of threshold value for the gradient of the filtered data, along with the Laplacian edge detector used for cross-checking, can be chosen to minimize false alarms. For the particular faults and noise levels considered in this chapter, only the low and high rotor speeds were found to be able to give zero false alarms and zero missed alarms. In general, measurements with less noise are more suited to trend detection. One of the benefits of the trend shift detection approach proposed in this chapter is that it will work regardless of whether the trend shift is from zero or from some other nonzero quasi-steady-state measurement value.

7

Optimally Weighted Recursive Median Filters

The previous chapters have showcased a number of signal processing approaches for gas turbine diagnostics. In general, the filters have become more complex and powerful from Chapter 2 to Chapter 6. The issue of non-Gaussian noise remains an important aspect for filters developed for fault detection. For example, Yoshida [71] points out that non-Gaussian noise occurs in health signals because damage tends to be concentrated in a specific part of the structure. He used a Monte Carlo filter to address the issue of non-Gaussian noise in structural damage detection for structures following earthquakes. However, the computer time requirements for such a filtering method can be very large.

We have seen in the earlier chapters that median filters can be used to preprocess health signals before subjecting them to fault detection and isolation algorithms. There is a possibility of significantly enhancing the median filters for diagnostics preprocessing applications. Progressive improvements in advancing median type algorithms were made over the last decade. In Chapter 6, the use of recursive median filters was studied, and it was found that such filters have excellent noise removal properties. Comparing wavelets with recursive median filters for denoising frequency time series for improved operational diagnostics, it was found that wavelets provide greater levels of noise reduction, and recursive median filters provide good results and are much simpler to develop and implement. Moreover, the non-linear nature of the median type filters makes them useful for the removal of outliers [66, 67].

Figure 1.1 shows a schematic of a gas turbine diagnostics system. We address the noise removal function in this chapter. Noise removal enhances both the automated and human-driven actions for diagnostics. In this chapter, the weighted recursive median filter is introduced for diagnostics applications. The concept of determining the optimal weights for different types of health signals is explored. A comprehensive study of this filter structure shows superior performance compared to the filters discussed earlier. The optimally weighted recursive median filters are the tools that can be of great use for denoising of signals before performing fault detection and isolation functions. This chapter is based on the research by Uday and Ganguli [72], who proposed the use of the optimally weighted recursive median filter for gas turbine diagnostics.

7.1 Weighted Recursive Median Filters

The performance of the recursive median filter can be greatly improved by the use of weights. This allows the filter to be tuned to particular types of signals and to reduce the blurring and streaking effects that are observed in the recursive median filter. Moreover, recasting the RM filter in this form provides faster implementation. The weighted recursive median filter can be represented as [38]

$$y_k = median(w_{k-n} \circ y_{k-n}, w_{k-n+1} \circ y_{k-n+1}, \cdots, w_k \circ x_k, \cdots, w_{k+n-1} \circ x_{k+n-1}, w_{k+n} \circ x_{k+n})$$

$$(7.1)$$

Here $\circ$ stands for duplication and w are the integer weights. Duplication implies that the data point is repeated. For example, $y_k = median$ $(2 \circ y_{k-1}, 3 \circ x_k, x_{k+1})$ is same as $y_k = median(y_{k-1}, y_{k-1}, x_k, x_k, x_k, x_{k+1})$. Weighted RM filters can be of two types based on the weights used:

1. Weighted symmetric recursive median filters
2. Weighted asymmetric recursive median filters

Symmetrically weighted filters are structures in which the weights are chosen to be symmetric, i.e., $w_{n-i} = w_{n+i}$ [73]. However, in the nonsymmetric structure, the weight values do not follow any particular pattern. Nonsymmetric filters may have advantages over symmetric filters and will be discussed later in this chapter. Signal processing literature has proposed adaptive approaches to weighting these filters based on mathematical methods. Such approaches are quite complicated and require a mathematical model of the system.

The primary focus of this chapter is to explore the possibilities provided by the weighted filters for noise reduction in health signals. We find the weights that offer the best denoising performance for typical health signals using an optimization approach.

7.2 Test Signals

In this chapter, an ideal root signal ΔEGT with implanted HPC or HPT faults is used to test the filters. The other root signals for $\Delta N1$, $\Delta N2$, and ΔWF can be similarly derived from the engine model [11].

Consider the basic measurement deltas ΔEGT, $\Delta N1$, $\Delta N2$, and ΔWF. In all practical applications, a certain level of noise is always present in the measured signal. As a result, these measurement deltas can be expressed as

$$z = z^0 + \varepsilon \tag{7.2}$$

where ε represents the noise, z^0 is the pure measurement delta, also called the root signal, and z is the noisy or corrupted signal. Hence, a filter Ψ is required to remove the noise and return the filtered signal for proper damage detection.

$$\hat{z} = \Psi(z) = \Psi(z^0 + \varepsilon) \tag{7.3}$$

For a comprehensive study of the role of weighted recursive median filters in eliminating noise from gas path measurements, four different signals are considered. These signals form the basic representation of the most common types of health signals:

1. Step signal (indicating an abrupt fault)
2. Ramp signal (indicating a gradual fault)
3. Combination signal (comprising both abrupt and gradual faults)
4. Transient gas path signal (obtained from [54], used in Chapter 5 with myriad filter)

While the first three signals simulate steady-state gas path measurements, the transient type signal can sometimes provide information about the engine that is not true in steady-state signals. Each of the first three signals comprises 200 data points. The root signal in Figure 7.1 depicts a step

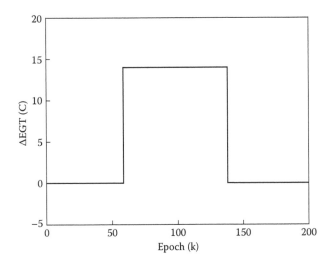

FIGURE 7.1
Step signal representing an HPC fault and its repair. (From Uday, P., and Ganguli, R., *Journal of Engineering for Gas Turbines and Power* 132(4):2010. With permission.)

signal, and it represents a single fault, which may be triggered by an event such as foreign object damage. Data point $k = 60$ represents the onset of this fault. The damage caused is identified as a 2% fall in HPC efficiency and the HPC module is repaired at point $k = 140$. In Figure 7.2, the development of the HPT fault is illustrated by use of the ramp signal. This fault differs from the HPC such that the fault does not occur suddenly; it develops due to engine deterioration. Again, the maximum value of ΔEGT here corresponds to a 2% fall in HPT efficiency. Here, the growth is gradual (approximated by a linear function) from points $k = 40 - 116$. From $k = 116$ the HPT fault remains steady and is finally repaired at $k = 140$. The step and ramp signals represent the two types of faults considered individually.

Figure 7.3 shows a combination signal, wherein both types of faults may occur one after the other. This is a more practical case since any gas turbine is susceptible to both faults. Figure 7.4 represents a transient signal of a deteriorated engine with a single-component fault in the intermediate-pressure compressor (IPC) implanted. This signal is obtained from [54] and is also used in [53].

In the ideal scenario, the measurement delta is clearly defined at each point. However, to efficiently test the performance of the filters, these signals have been subjected to certain noise levels by using additive white Gaussian noise with varying signal-to-noise ratios (SNRs). Hence, the performance of weighted recursive median filters is studied for high-noise (SNR = 0.1), medium-noise (SNR = 0.3), and low-noise (SNR = 1.5) signals.

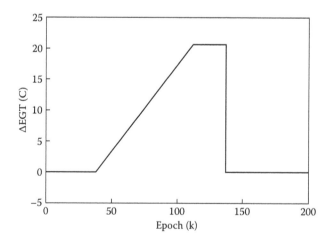

FIGURE 7.2
Ramp signal representing an HPT fault and its repair. (From Uday, P., and Ganguli, R., *Journal of Engineering for Gas Turbines and Power* 132(4):2010. With permission.)

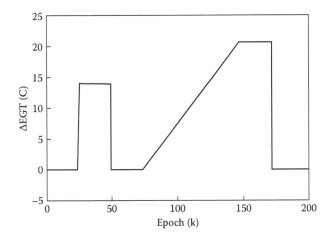

FIGURE 7.3
Combination signal (step and ramp) representing an HPC fault and its repair followed by an HPT fault and its repair. (From Uday, P., and Ganguli, R., *Journal of Engineering for Gas Turbines and Power* 132(4):2010. With permission.)

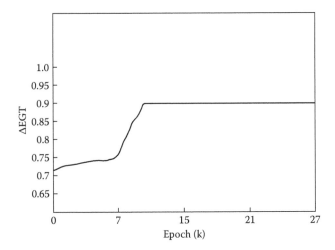

FIGURE 7.4
Transient gas path signal representing IPC fault and transient data. (From Uday, P., and Ganguli, R., *Journal of Engineering for Gas Turbines and Power* 132(4):2010. With permission.)

7.3 Numerical Simulations

The weighted RM filter is first compared to the finite impulse response (FIR), infinite impulse response (IIR), and simple median filter, and then to the traditional or unweighted RM filter. The recursive median filter used in this chapter is a five-point filter with no weights, and the weighted recursive median filters used here were discussed earlier in Equation (7.1). To obtain a quantitative idea of the noise reduction, we look at two types of error criteria. The root mean square (*RMS*) error is a measure of the difference between the filtered and the ideal signal. This is given as

$$RMS = \sqrt{\frac{1}{N}\sum_{i=1}^{N}(\Delta\hat{z}_i - \Delta z_i^0)^2} \qquad (7.4)$$

Here N is the number of data points in the sample. The minimum absolute error (*MAE*), which is more sensitive to outliers, is also used to test the filters [38]. We will add outliers to the test signals later in the chapter. In the *MAE* criterion, the error is defined as

$$MAE = \sum_{i=1}^{N}\frac{1}{N}\left|\Delta\hat{z}_i - \Delta z_i^0\right| \qquad (7.5)$$

Tables 7.1 and 7.2 summarize the results obtained on passing the test signals through the different filters for the mean *RMS* and *MAE* estimates, respectively.

Since random noise is added to the signals, each noisy signal is different and a large number of such signals should be filtered to arrive at an accurate estimate of the noise reduction given by the filter. Therefore, we use 1000 samples of noisy data to arrive at the mean *RMS* and *MAE* values. We see that the RM filter performs better than the SM filter for all cases.

We now consider the five-point weighted median filter defined as

$$y_k = median\left(w_{-2}\circ y_{k-2}, w_{-1}\circ y_{k-1}, w_0\circ x_k, w_1\circ x_{k+1}, w_2\circ x_{k+2}\right) \qquad (7.6)$$

The filter in Equation (7.6) has five integer weights. The five-point filter keeps the time delay to only two points since x_{k+1} and x_{k+2} are needed by the filter. For many engines, data are available at a few points during each flight. Therefore, the low filter length keeps the time delay to a minimum while providing sufficient filter length for noise removal, say, in comparison to a three-point filter. If data are available more rapidly, longer-length filters

TABLE 7.1

Mean *RMS* Error Estimates of Five-Point Filters on Test Signals

Signal Type	SNR Value	Simple Median (SM)	Recursive Median (RM)	Weighted Recursive Median (WRM)
	0.1	0.5441	0.4678	0.3806
Step	0.3	0.5327	0.4532	0.3731
	1.5	0.462	0.3974	0.3242
	0.1	0.5602	0.5594	0.4554
Ramp	0.3	0.5581	0.5475	0.4481
	1.5	0.4989	0.4889	0.399
	0.1	0.5971	0.594	0.4911
Combination	0.3	0.5848	0.5738	0.4826
	1.5	0.529	0.5134	0.4099
	0.1	0.5287	0.4352	0.3446
Transient signal	0.3	0.5187	0.4243	0.3376
	1.5	0.4509	0.3764	0.2944

Source: Uday, P., and Ganguli, R., *Journal of Engineering for Gas Turbines and Power* 132(4):2010. With permission.

TABLE 7.2

Mean *MAE* of *WRM* Filter of Different Window Length on Test Signals

Signal Type	SNR Value	Simple Median (SM)	Recursive Median (RM)	Weighted Recursive Median (WRM)
	0.1	0.4277	0.3576	0.2872
Step	0.3	0.4190	0.3459	0.2806
	1.5	0.3638	0.3031	0.2428
	0.1	0.4460	0.4311	0.3506
Ramp	0.3	0.4268	0.4210	0.3444
	1.5	0.3856	0.3739	0.3054
	0.1	0.4660	0.4560	0.3790
Combination	0.3	0.4503	0.4403	0.3728
	1.5	0.3999	0.3930	0.3300
	0.1	0.4227	0.3454	0.2736
Transient signal	0.3	0.4156	0.3373	0.2664
	1.5	0.3608	0.2982	0.2328

Source: Uday, P., and Ganguli, R., *Journal of Engineering for Gas Turbines and Power* 132(4):2010. With permission.

can be considered. To obtain the optimal weights we solve the following optimization problem. Minimize

$$f(w_{-2}, w_{-1}, w_0, w_1, w_2) = \frac{\sum_{i=1}^{M} RMS}{M} \tag{7.7}$$

Here $M = 1000$ samples of noisy data are used to obtain a mean RMS error, and the weights are design variables of the filter that need to be determined for minimum error. For applications with the weighted filter, all combinations of design variables or weights are computed using integer values {1, 2, 3, and 4}. It is found that using higher weights yields the same filter as lower weights because of duplication in the median operation. For example, the weights (4, 1, 3, 2, and 4) give the same result as the weights (8, 2, 6, 4, and 8) in terms of median value. However, the lower weight set is more efficient. Through exhaustive numerical search of the design space, it is observed that several groups of weights could be used to reduce the mean RMS error to below that produced by the standard recursive median filter. For computer implementation, the weighted recursive median filter is placed inside a loop of 1000 iterations to obtain the average reduction in noise for a given weight set. This loop is then placed inside a nested loop of depth 5, which varies the weights from 1 to 4 in intervals of 1. Thus, the noise removal in terms of RMS error for all the integer weights is obtained and the weights corresponding to the minimum values of RMS error are selected. A similar exercise is performed using the MAE criteria by minimizing the objective function

$$f(w_{-2}, w_{-1}, w_0, w_1, w_2) = \frac{\sum_{i=1}^{M} MAE}{M} \tag{7.8}$$

Here $M = 1000$ samples of noisy data are used to obtain a mean MAE and the weights are design variables of the filter that need to be determined for minimum error. The optimum set of weights is arrived at by determining the lowest RMS and MAE error values that could be achieved for each signal. These optimum weights are shown in Table 7.3 for the RMS error and the MAE error. The same set of weights gives both the lowest RMS and MAE errors in these cases.

The weighted recursive median (WRM) filter results in Tables 7.1 and 7.2 correspond to the optimal weights in Table 7.3. We see that the WRM filter shows a significant improvement in noise removal compared to the other filters. An interesting observation is the lack of one universal sequence of weights that minimizes the error. This implies that there exists an exclusive group of weights for each signal that can completely minimize errors. Most general steady-state signals will be similar to the combination signal, and therefore the weight set (2, 2, 2, 1, 3) can be used for such signals.

TABLE 7.3

Optimal Weights for Five-Point Weighted Recursive
Median (WRM) Filter Using Both *RMS* and *MAE* Criteria

Signal Type	SNR Value	Weights
	0.1	[4,1,3,2,4]
Step	0.3	[4,1,3,2,4]
	1.5	[4,1,3,2,4]
	0.1	[2,1,2,1,2]
Ramp	0.3	[2,1,2,1,2]
	1.5	[2,1,2,1,2]
	0.1	[2,2,2,1,3]
Combination	0.3	[2,2,2,1,3]
	1.5	[2,2,2,1,3]
	0.1	[4,1,3,2,4]
Transient signal	0.3	[4,1,3,2,4]
	1.5	[4,1,3,2,4]

Source: Uday, P., and Ganguli, R., *Journal of Engineering for Gas Turbines and Power* 132(4):2010. With permission.

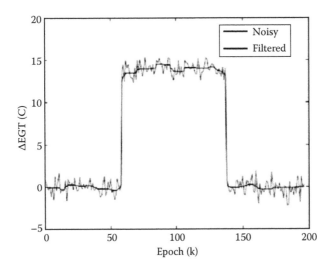

FIGURE 7.5
Effect of weighted RM filters on noisy step signal with SNR = 1.5. (From Uday, P., and Ganguli, R., *Journal of Engineering for Gas Turbines and Power* 132(4):2010. With permission.)

On the other hand, the weights (4, 1, 2, 3, 4) can be used for the transient signals. Physically, the weights mean that certain samples in the signal are given more importance than others. The weights are sensitive to the signal type rather than to noise levels. Figures 7.5–7.8 visually represent the effects of the weighted recursive median filter on the test signals with SNR of 1.5.

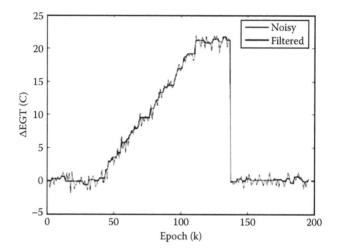

FIGURE 7.6
Effect of weighted RM filters on noisy ramp signal with SNR = 1.5. (From Uday, P., and Ganguli, R., *Journal of Engineering for Gas Turbines and Power* 132(4):2010. With permission.)

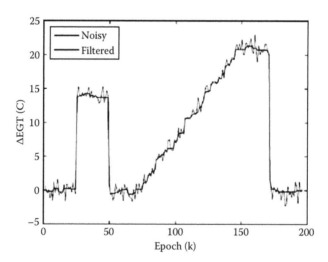

FIGURE 7.7
Effect of weighted RM filters on noisy combination signal with SNR = 1.5. (From Uday, P., and Ganguli, R., *Journal of Engineering for Gas Turbines and Power* 132(4):2010. With permission.)

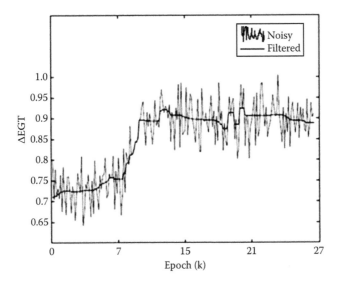

FIGURE 7.8
Effect of weighted RM filters on noisy realistic signal with SNR = 1.5. (From Uday, P., and Ganguli, R., *Journal of Engineering for Gas Turbines and Power* 132(4):2010. With permission.)

These figures clearly illustrate the capability of the weighted RM filters to preserve sharp edges or trend shifts in a signal and to remove noise from stationary regions.

7.4 Test Signal with Outliers

The signal in Figure 7.9 considers the combination signal with added noise (SNR = 1.5) and outliers. Outliers represent the impulsive noise that may be present in a signal. Here, the outliers are selected at three different levels. The first is equal to 4.23°C and is added at $k = 10$, 80, and 140 and subtracted at $k = 40$ and 120. The 8.46°C outlier is added at $k = 20$, 100, and 190 and subtracted at $k = 30$ and 170. The last outlier has a value of 12.69°C, and this is added at $k = 110$ and 160 and subtracted at $k = 60$ and 130. Similarly, outliers are added to the step, ramp, and transient signals. The weights obtained after putting in the outliers are the same as shown in Table 7.3. This is primarily because all median architectures are good at removing outliers, and the weights serve to address the ideal signal characteristics.

The weighted recursive median filter is able to efficiently discard these outliers while preserving signal features that can be easily observed from Figures 7.9–7.12. Results in Tables 7.4 and 7.5 show that the simple median

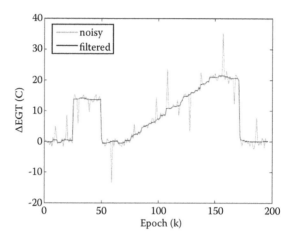

FIGURE 7.9
Effect of weighted RM filters on noisy combination signal with outliers. (From Uday, P., and Ganguli, R., *Journal of Engineering for Gas Turbines and Power* 132(4):2010. With permission.)

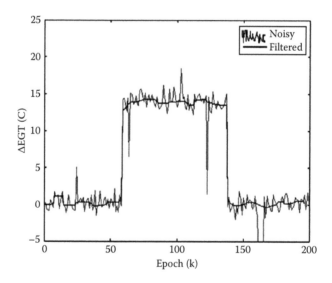

FIGURE 7.10
Effect of weighted RM filters on noisy step signal with outliers. (From Uday, P., and Ganguli, R., *Journal of Engineering for Gas Turbines and Power* 132(4):2010. With permission.)

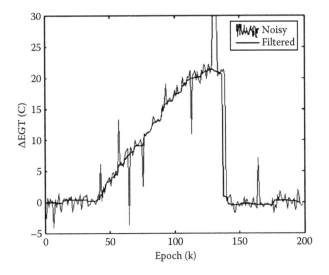

FIGURE 7.11
Effect of weighted RM filters on noisy ramp signal with outliers. (From Uday, P., and Ganguli, R., *Journal of Engineering for Gas Turbines and Power* 132(4):2010. With permission.)

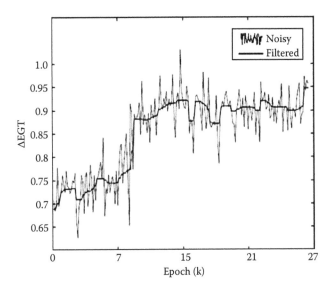

FIGURE 7.12
Effect of weighted RM filters on noisy realistic signal with outliers. (From Uday, P., and Ganguli, R., *Journal of Engineering for Gas Turbines and Power* 132(4):2010. With permission.)

TABLE 7.4

RMS Error of Different Filters on Test Signal Containing Outliers

Signal Type	SNR Value	Simple Median (SM)	Recursive Median (RM)	Weighted Recursive Median (WRM)
Step	0.1	0.5632	0.4819	0.3977
	0.3	0.5523	0.4750	0.3866
	1.5	0.4788	0.4133	0.3424
Ramp	0.1	0.6016	0.5730	0.4922
	0.3	0.5900	0.5646	0.4795
	1.5	0.5272	0.4961	0.4311
Combination	0.1	0.6332	0.6113	0.5291
	0.3	0.6238	0.5994	0.5183
	1.5	0.5500	0.5249	0.4632
Transient signal	0.1	0.5292	0.4372	0.3392
	0.3	0.5182	0.4281	0.3362
	1.5	0.4502	0.3752	0.2968

Source: Uday, P., and Ganguli, R., *Journal of Engineering for Gas Turbines and Power* 132(4):2010. With permission.

TABLE 7.5

MAE Estimate of Different Filters on Test Signal Containing Outliers

Signal Type	SNR Value	Simple Median (SM)	Recursive Median (RM)	Weighted Recursive Median (WRM)
Step	0.1	0.4422	0.3666	0.2998
	0.3	0.4346	0.3618	0.2918
	1.5	0.3761	0.3156	0.2570
Ramp	0.1	0.4594	0.4511	0.3754
	0.3	0.4495	0.4439	0.3667
	1.5	0.3996	0.3897	0.3275
Combination	0.1	0.4850	0.4791	0.4102
	0.3	0.4774	0.4692	0.4011
	1.5	0.4200	0.4112	0.3576
Transient signal	0.1	0.4249	0.3466	0.2705
	0.3	0.4155	0.3397	0.2671
	1.5	0.3610	0.2967	0.2354

Source: Uday, P., and Ganguli, R., *Journal of Engineering for Gas Turbines and Power* 132(4):2010. With permission.

and RM filters do not provide the same degree of immunity to noise and outliers. This superior performance of the weighted filter makes it highly suitable for denoising of engine health signals, where the features of the original signal are critical to engine maintenance.

7.5 Performance Comparison

This section summarizes the filter performance of the structures used in this chapter. To get a statistical measure of the effectiveness of each method, 1000 samples of noisy data are generated and filtered as discussed previously. Using the *MAE* definition, as given in Equation (7.5), we can define a parameter for efficiency measurement of these filters in terms of noise reduction as

$$\rho = \frac{MAE^{(noisy)} - MAE^{(filtered)}}{MAE^{(noisy)}} 100 \qquad (7.9)$$

Table 7.6 clearly illustrates this improvement by comparing the performance of the three filters studied in this work.

We observe that the simple median filter provides a noise reduction of only 39 to 46%; the traditional recursive structure brings about a reduction of 41 to 56%, while the weighted structure improves this considerably to values ranging from 51 to 65%. This leads to significant accuracy in obtaining the root signal from the contaminated one.

TABLE 7.6

Percentage Noise Reduction Provided by Different Filters for Test Signals

Signal Type	SNR Value	$\rho^{(Median)}$%	$\rho^{(RMF)}$%	$\rho^{(Weighted\ RMF)}$%
	0.1	45.63	54.54	63.49
Step	0.3	45.54	55.04	63.53
	1.5	45.84	54.90	63.87
	0.1	43.34	45.48	55.46
Ramp	0.3	44.73	45.48	55.40
	1.5	42.64	44.38	54.57
	0.1	40.93	42.20	51.96
Combination	0.3	41.53	42.80	51.62
	1.5	39.06	41.57	50.94
	0.1	46.55	56.32	65.40
Transient signal	0.3	46.03	56.20	65.41
	1.5	46.18	55.52	65.27

Source: Uday, P., and Ganguli, R., *Journal of Engineering for Gas Turbines and Power* 132(4):2010. With permission.

In order to clearly observe the superiority of the weighted filter, we use another parameter, η, which is defined as

$$\eta^{(SM)} = \frac{MAE^{(SM)} - MAE^{(WRMF)}}{MAE^{(SM)}} 100 \qquad (7.10)$$

$$\eta^{(RMF)} = \frac{MAE^{(RMF)} - MAE^{(WRMF)}}{MAE^{(RMF)}} 100 \qquad (7.11)$$

Table 7.7 summarizes the improvement provided by the weighted filters over the simple median and recursive median filters. It is seen that weighted RM filters improve the effectiveness of the median filters by a maximum of 36%, and of the unweighted RM filters by 22%. These results prove that the weighted filter provides several advantages over the standard filters in terms of feature preservation and noise reduction. Note that these advantages are provided by a numerically efficient filter that is very easy to implement. For the numerical results, the filters are implemented in MATLAB®. The results of this chapter clearly show that optimization can lead to a better filter for diagnostic systems with negligible increase in complexity and cost. Tables 7.8 and 7.9 show the improvements in performance of the weighted filter over the other filters for signals that are contaminated with Gaussian noise as well as non-Gaussian outliers. From Table 7.8, we can see that the simple median filter reduces noise by about 46–65%, the recursive median filter by about 56–66% and the WRM filter by about 65–70%.

TABLE 7.7

Improvement in Performance of Weighted RM Filters over Other Filters

Signal Type	SNR Value	$\eta^{(SM)}$ (%)	$\eta^{(RMF)}$ (%)
	0.1	32.85	19.69
Step	0.3	33.03	18.88
	1.5	33.26	19.89
	0.1	21.39	18.67
Ramp	0.3	19.31	18.19
	1.5	20.80	18.32
	0.1	18.67	16.89
Combination	0.3	17.26	15.38
	1.5	19.49	16.03
	0.1	35.27	20.29
Transient signal	0.3	35.90	21.02
	1.5	35.48	21.93

Source: Uday, P., and Ganguli, R., *Journal of Engineering for Gas Turbines and Power* 132(4):2010. With permission.

TABLE 7.8

Percentage Noise Reduction Provided by Different Filters for Noisy Test Signal Contaminated with Outliers

Signal Type	SNR Value	$\rho^{(Median)}$%	$\rho^{(RMF)}$%	$\rho^{(Weighted\ RMF)}$%
Step	0.1	51.84	60.07	67.35
	0.3	51.90	59.96	67.70
	1.5	53.17	60.70	68.00
Ramp	0.1	57.16	57.93	64.99
	0.3	57.39	57.92	65.24
	1.5	58.24	59.27	65.77
Combination	0.1	62.85	63.30	68.58
	0.3	62.92	63.56	68.85
	1.5	64.83	65.57	70.06
Transient signal	0.1	46.16	56.08	65.72
	0.3	46.14	55.97	65.38
	1.5	46.16	55.75	64.89

Source: Uday, P., and Ganguli, R., *Journal of Engineering for Gas Turbines and Power* 132(4):2010. With permission.

TABLE 7.9

Improvement in Performance of Weighted RM Filters over Other Filters for Noisy Test Signals Contaminated with Outliers

Signal Type	SNR Value	$\eta^{(SM)}$ (%)	$\eta^{(RMF)}$ (%)
Step	0.1	32.30	18.22
	0.3	32.86	19.35
	1.5	31.67	18.57
Ramp	0.1	18.28	16.78
	0.3	18.42	17.39
	1.5	18.04	15.96
Combination	0.1	15.42	14.38
	0.3	15.98	14.51
	1.5	14.86	13.04
Transient signal	0.1	36.34	21.96
	0.3	35.82	21.37
	1.5	34.79	20.66

Source: Uday, P., and Ganguli, R., *Journal of Engineering for Gas Turbines and Power* 132(4):2010. With permission.

7.6 Three- and Seven-Point Optimally Weighted RM Filters

In this section, the effect of integer weight on the three-point WRM filter and the seven-point WRM filter is shown. Optimal weights for different health signals are obtained. The three-point filter is useful for turbines where data are obtained slowly, and the seven-point filter where data are available rapidly. The effect of window length on the optimal weights of WRM filters developed for gas turbine diagnostics is brought out. The idea of using a three- and seven-point WRM filter was proposed by Guruprakash and Ganguli [74] and is discussed in this section.

7.6.1 Numerical Analysis

To obtain the optimal weights we solve the following optimization problem; for the three-point WRM filter $y_k = median(w_{-1} \circ y_{-1}, w_0 \circ x_0, w_1 \circ x_1)$ the optimization problem is to minimize

$$f(w_{-1}, w_0, w_1) = \frac{\sum_{i=1}^{M} RMS}{M} \tag{7.12}$$

where $w_{-1}, w_0,$ and w_1 are the three integer design variables.

For the seven-point filter $y_k = median(w_{-3} \circ y_{-3}, w_{-2} \circ y_{-2}, w_{-1} \circ y_{-1}, w_0 \circ x_0, w_1 \circ x_1, w_2 \circ x_2, w_3 \circ x_3)$, we need to minimize

$$f(w_{-3}, w_{-2}, w_{-1}, w_0, w_1, w_2, w_3) = \frac{\sum_{i=1}^{M} RMS}{M} \tag{7.13}$$

where $w_{-3}, w_{-2}, w_{-1}, w_0, w_1, w_2, w_3$ are the seven integer design variables. Here, $M = 1000$ samples of noisy data are simulated to obtain a mean RMS error, and the weights are design variables that are obtained for minimum error. It can be observed that the three-point filter has a time lag of 1 (it requires a forward data point x_1) and the seven-point filter has a time lag of 3 (it needs x_1, x_2, x_3). The five-point filter $y_k = median(w_{-2} \circ y_{-2}, w_{-1} \circ y_{-1}, w_0 \circ x_0, w_1 \circ x_1, w_2 \circ x_2)$ has a time lag of 2 (it needs x_1 and x_2) [72].

For many turbines, only a few data points are available per flight. So the three-point filter is suitable for trend monitoring and diagnostics of such turbines. Some recent engines have higher rates of data acquisition. For such turbines, the seven-point filter is useful. For application with the weighted filter, all combinations of weights are computed using integer values {1, 2, 3, 4}. Through exhaustive numerical search of the design space, it is observed that several groups of weights could be used to reduce the RMS error.

TABLE 7.10

Mean *RMS* Error of WRM Filter of Different Window Length on Test Signals

Signal Type	SNR Value	Three-Point WRM Filter	Five-Point WRM Filter	Seven-Point WRM Filter
	0.1	0.5573	0.3806	0.3689
Step	0.3	0.5208	0.3731	0.3597
	1.5	0.5019	0.3242	0.3166
	0.1	0.6854	0.4554	0.4331
Ramp	0.3	0.6547	0.4481	0.4232
	1.5	0.6193	0.3990	0.4109
	0.1	0.7157	0.4911	0.4813
Combination	0.3	0.6994	0.4826	0.4698
	1.5	0.6279	0.4099	0.3988
	0.1	0.5468	0.3446	0.3118
Transient signal	0.3	0.5293	0.3376	0.3042
	1.5	0.5816	0.2944	0.2758

Source: Guruprakash, V.N., and Ganguli, R., *ASME Journal of Engineering in Gas Turbine and Power* 133(10): article 104502, 2011. With permission.

Thus, weights corresponding to minimum values of *RMS* error are selected. Table 7.10 shows the *RMS* error of three-point WRM and seven-point WRM filters, along with results of the five-point filter from [72]. We can see that the error reduces as the window length increases.

Similarly, *MAE* criteria with objective functions are used to minimize the error of three-point and seven-point filters. For the three-point filter, we minimize

$$f(w_{-1}, w_0, w_1) = \frac{\sum_{i=1}^{M} MAE}{M} \qquad (7.14)$$

For the seven-point filter, the objective function is

$$f(w_{-3}, w_{-2}, w_{-1}, w_0, w_1, w_2, w_3) = \frac{\sum_{i=1}^{M} MAE}{M} \qquad (7.15)$$

Here, $M = 1000$ samples of noisy data are simulated to obtain a mean *MAE* and the weights are design variables that are obtained for minimum error. Table 7.11 shows the *MAE* of the three-point WRM filter and the seven-point WRM filter along with the five-point filter.

Optimal sets of weights are obtained by considering the minimum error from *RMS* and *MAE* criteria for both three-point WRM and seven-point WRM filters. Table 7.12 shows the optimal set of weights for three-point and

TABLE 7.11

Mean *MAE* of WRM Filter of Different Window Length on Test Signals

Signal Type	SNR Value	Three-Point WRM Filter	Five-Point WRM Filter	Seven-Point WRM Filter
	0.1	0.4658	0.2872	0.2714
Step	0.3	0.4291	0.2806	0.2697
	1.5	0.4098	0.2428	0.2388
	0.1	0.5487	0.3506	0.3447
Ramp	0.3	0.5312	0.3444	0.3379
	1.5	0.5149	0.3054	0.3017
	0.1	0.5913	0.3790	0.3682
Combination	0.3	0.5839	0.3728	0.3602
	1.5	0.5551	0.3300	0.3025
	0.1	0.4982	0.2736	0.2604
Transient signal	0.3	0.4569	0.2664	0.2418
	1.5	0.4213	0.2328	0.2114

Source: Guruprakash, V.N., and Ganguli, R., *ASME Journal of Engineering in Gas Turbine and Power* 133(10): article 104502, 2011. With permission.

TABLE 7.12

Optimal Set of Weights for Three-Point and Seven-Point WRM Filter

Signal Type	SNR Value	Three-Point WRM Filter	Five-Point WRM Filter	Seven-Point WRM Filter
	0.1	[4 2 2]	[4 1 3 2 4]	[4 3 3 2 2 2 1]
Step	0.3	[4 2 2]	[4 1 3 2 4]	[4 3 3 2 2 2 1]
	1.5	[4 2 2]	[4 1 3 2 4]	[4 3 3 2 2 2 1]
	0.1	[4 3 3]	[2 1 2 1 2]	[4 2 4 2 1 4 2]
Ramp	0.3	[4 3 3]	[2 1 2 1 2]	[4 2 4 2 1 4 2]
	1.5	[4 3 3]	[2 1 2 1 2]	[4 2 4 2 1 4 2]
	0.1	[4 4 3]	[2 2 2 1 3]	[4 3 2 2 2 1 4]
Combination	0.3	[4 4 3]	[2 2 2 1 3]	[4 3 2 2 2 1 4]
	1.5	[4 4 3]	[2 2 2 1 3]	[4 3 2 2 2 1 4]
	0.1	[4 1 4]	[4 1 3 2 4]	[4 4 2 3 1 4 2]
Transient signal	0.3	[4 1 4]	[4 1 3 2 4]	[4 4 2 3 1 4 2]
	1.5	[4 1 4]	[4 1 3 2 4]	[4 4 2 3 1 4 2]

Source: Guruprakash, V.N., and Ganguli, R., *ASME Journal of Engineering in Gas Turbine and Power* 133(10): article 104502, 2011. With permission.

seven-point WRM filters, respectively. For a given signal type and filter length, the optimal weights obtained using both *RMS* and *MAE* measures remain the same, even though the signal-to-noise ratios change. Figures 7.13–7.16 show the filtering effect of the three-point WRM filter. Figures 7.17–7.20 show the filtering effect of the seven-point WRM filter. It can be observed that

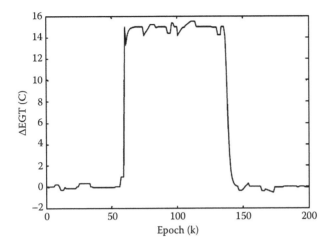

FIGURE 7.13
Effect of three-point WRM filter on noisy step signal with SNR = 1.5.

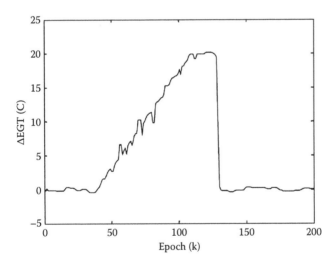

FIGURE 7.14
Effect of three-point WRM filter on noisy ramp signal with SNR = 1.5.

the seven-point filter is a good smoother. However, even the three-point filter is able to remove a substantial amount of noise from the signal.

7.6.2 Signal with Outliers

To further evaluate the performance of three-point WRM and seven-point WRM filters, the combination signal is used as a test signal with SNR = 1.5,

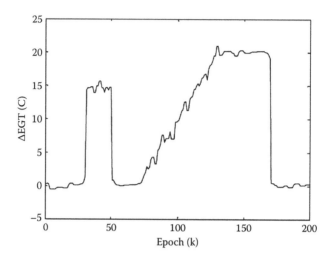

FIGURE 7.15
Effect of three-point WRM filter on noisy combined signal with SNR = 1.5.

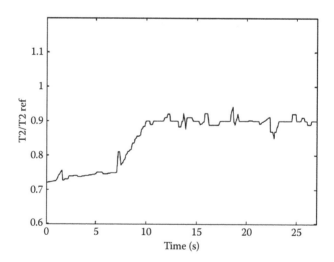

FIGURE 7.16
Effect of three-point WRM filter on noisy transient signal with SNR = 1.5.

and outliers are considered at three different levels, as shown in Figure 7.21. The first level is equal to 4.23°C and is added at $k = 10$, 80, and 140 and subtracted at $k = 40$ and 120. The next level is an 8.46°C outlier that is added at $k = 20$, 100, and 190 and subtracted at $k = 30$ and 170. The last level has a value of 12.69°C, and this is added at $k = 110$ and 160 and subtracted at $k = 60$ and 130. Outliers are added to the rest of the test signals in a similar way and are shown in Figures 7.22–7.24. The effect of the three-point

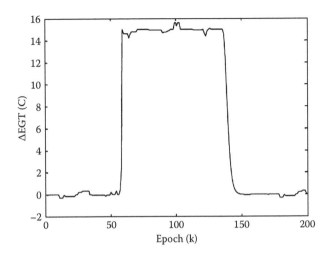

FIGURE 7.17
Effect of seven-point WRM filter on noisy step signal with SNR = 1.5.

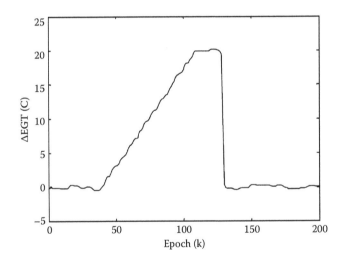

FIGURE 7.18
Effect of seven-point WRM filter on noisy ramp signal with SNR = 1.5.

WRM filter on signals with outliers is shown in Figures 7.25–7.28. Filtered signals that are passed through the seven-point WRM signal are shown in Figures 7.29–7.32.

The three-point and seven-point WRM filters effectively attenuate the outliers from the test signals, keeping the signal feature unchanged. As expected, the long window seven-point filters show better performance. Table 7.13 shows the *RMS* error of three-point WRM and seven-point WRM filters with

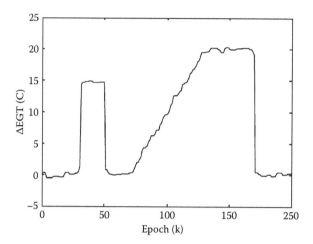

FIGURE 7.19
Effect of seven-point WRM filter on noisy combined signal with SNR = 1.5.

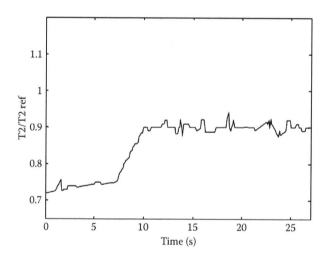

FIGURE 7.20
Effect of seven-point WRM filter on noisy transient signal with SNR = 1.5.

an outlier in the test signal. The error for the five-point WRM is also shown in the table. It can be observed that the error decreases as the filter window length increases.

Table 7.14 shows the *MAE* of three-point WRM and seven-point WRM filters with an outlier in the test signal, along with the five-point results. It can be seen that the filters perform well and are good tools for preprocessing

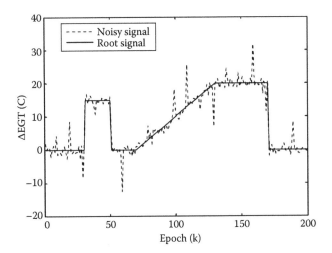

FIGURE 7.21
Combination signal with noise and outliers.

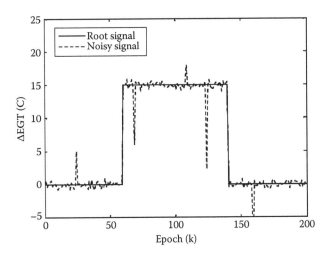

FIGURE 7.22
Step signal with noise and outliers.

gas turbine measurement deltas before using them for fault detection and isolation.

We have shown that optimally weighted recursive median architectures are powerful tools for improved gas turbine diagnostics. An adaptive approach to weight generation based on incoming online data can be explored. Other types of faults with different magnitudes (both bias and slope) should be

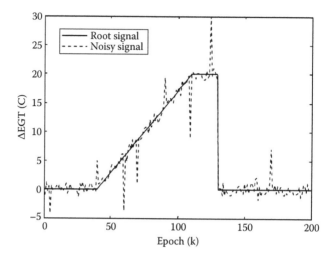

FIGURE 7.23
Ramp signal with noise and outliers.

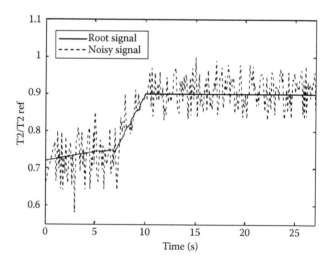

FIGURE 7.24
Transient signal with noise and outliers.

evaluated. Besides the abrupt faults and gradual faults considered in this chapter, other faults, such as intermittent faults and increase of noise caused by faults, are also possible and need to be addressed. We also note that the current approach can be used as a complement and preprocessor to other gas path denoising approaches [75].

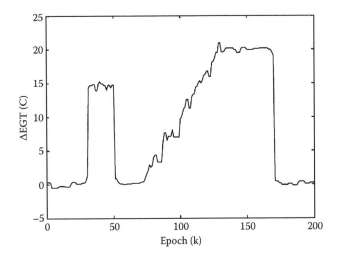

FIGURE 7.25
Combination signal including outliers after filtering with three-point WRM filter.

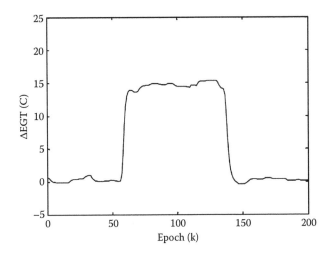

FIGURE 7.26
Step signal including outliers after filtering with three-point WRM filter.

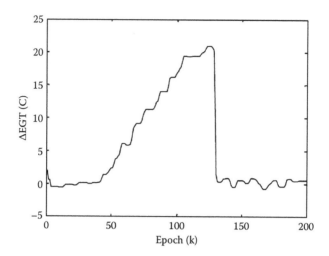

FIGURE 7.27
Ramp signal including outliers after filtering with three-point WRM filter.

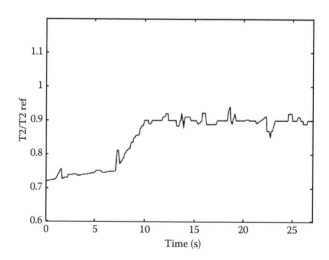

FIGURE 7.28
Transient signal including outliers after filtering with three-point WRM filter. (From Guruprakash, V.N., and Ganguli, R., *ASME Journal of Engineering in Gas Turbine and Power* 133(10): article 104502, 2011. With permission.)

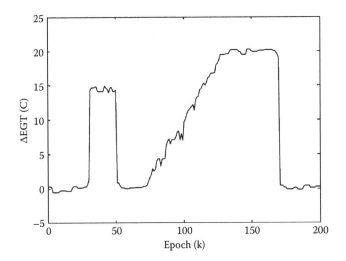

FIGURE 7.29
Combination signal including outliers after filtering with seven-point WRM filter.

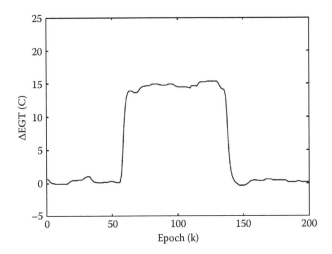

FIGURE 7.30
Step signal including outliers after filtering with seven-point WRM filter.

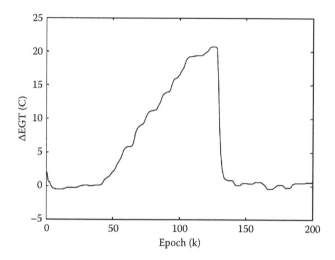

FIGURE 7.31
Ramp signal including outliers after filtering with seven-point WRM filter.

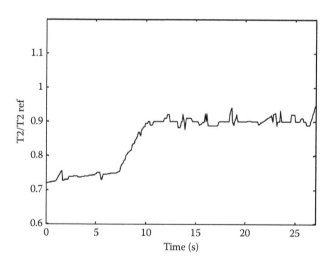

FIGURE 7.32
Transient signal including outliers after filtering with seven-point WRM filter. (From Guruprakash, V.N., and Ganguli, R., *ASME Journal of Engineering in Gas Turbine and Power* 133(10): article 104502, 2011. With permission.)

TABLE 7.13

RMS Error of WRM Filter of Different Window Length on Test Signal with Outliers

Signal Type	SNR Value	Three-Point WRM Filter	Five-Point WRM Filter	Seven-Point WRM Filter
	0.1	0.6164	0.3977	0.3689
Step	0.3	0.5951	0.3866	0.3597
	1.5	0.5782	0.3424	0.3166
	0.1	0.7372	0.4922	0.4331
Ramp	0.3	0.6991	0.4795	0.4232
	1.5	0.6783	0.4311	0.4109
	0.1	0.8341	0.5291	0.4813
Combination	0.3	0.7847	0.5183	0.4698
	1.5	0.7019	0.4632	0.3988
	0.1	0.6547	0.3392	0.3118
Transient signal	0.3	0.6330	0.3362	0.3042
	1.5	0.5958	0.2968	0.2758

TABLE 7.14

MAE Error of WRM Filter of Different Window Length on Test Signal with Outliers

Signal Type	SNR Value	Three-Point WRM Filter	Five-Point WRM Filter	Seven-Point WRM Filter
	0.1	0.5147	0.2998	0.2897
Step	0.3	0.4995	0.2918	0.2824
	1.5	0.4724	0.2570	0.2488
	0.1	0.6012	0.3754	0.3604
Ramp	0.3	0.5854	0.3667	0.3541
	1.5	0.5664	0.3275	0.3114
	0.1	0.6447	0.4102	0.4005
Combination	0.3	0.6289	0.4011	0.3987
	1.5	0.6010	0.3576	0.3419
	0.1	0.5625	0.2705	0.2671
Transient signal	0.3	0.5302	0.2671	0.2593
	1.5	0.5117	0.2354	0.2287

7.7 Summary

In this chapter, a new optimally weighted recursive median (WRM) filter for denoising health signals is proposed. Test signals for abrupt and gradual faults are used for a gas turbine engine diagnostic problem, along with a transient signal. A five-point WRM filter is developed, and the weights are optimized for typical diagnostic signals by minimizing the error norms between the noisy and root signals. The WRM filter provides better denoising results than the simple median filter and recursive median filter. The WRM filter also improves the visual quality of the signals by removing the noise and outliers while preserving important features of the root signal, such as sharp edges and gradual shifts. The WRM filter is presented as a preprocessor for denoising health signals prior to fault detection and isolation in gas turbine engines.

The optimal integer weights of three- and seven-point weighted recursive median filters are also obtained for signals representative of gas turbine engine faults. Objective functions based on the root mean square and minimum absolute error criteria are created and used for the optimization problem. The design variables are the integer weights of the filter. Fortunately, the optimal weights are the same for a given signal type and filter length, even when the signal-to-noise ratio is varied. It is found that the filters perform well with noisy data and with data containing non-Gaussian outliers. The three-point filter is useful for many engines, as it has a limited one-point time delay that is good for situations where data are obtained slowly. The seven-point filter is useful for engines where data are obtained rapidly. Increasing the window length reduces the error given by the filter. Both filters are good preprocessing tools for gas turbine diagnostics.

Further chapters of the book focus on fault isolation. It should be noted that the filters presented in Chapters 2–7 can be hybridized in various ways, depending on the applications.

8

Kalman Filter

In the previous chapters, we focused on filters for removing non-Gaussian noise and outliers from gas turbine signals. The present chapter showcases the Kalman filter for fault detection and isolation techniques in gas turbines. The goal of gas turbine performance diagnostics is to accurately detect, isolate, and assess the changes in engine module performance, engine system malfunctions, and instrumentation problems from knowledge of measured parameters taken along the engine's gas path. Discernable shifts in engine speed, temperature, pressure, fuel flow, etc., provide the requisite information for determining the underlying shift in engine operation from a presumed nominal state. Historically, this type of analysis was performed through the use of a Kalman filter or one of its derivatives to simultaneously estimate several engine faults. The present chapter outlines the Kalman filter methodology, its relative merits and weaknesses. Some basic background on the Kalman filter and the related weighted least-squares approach have been provided in Chapter 1. The typical use of the Kalman filter in Chapter 1 was for gas path analysis, which is also called multiple-fault analysis. In this approach, the Kalman filter distributes the measurement shifts among a variety of module efficiencies, flow capacities, and areas. In this chapter, the Kalman filter will be used to solve the problem of isolating a single fault to the component level. The single faults under consideration include the engine modules, engine system, and instrumentation faults. The use of the Kalman filter as a single-fault isolator was proposed by Volponi, Depold, Ganguli, and Daguang, and the present chapter is largely based on this paper [12]. Furthermore, the use of the Kalman filter for sensor fault estimation will also be discussed in this chapter.

8.1 Kalman Filter Approach

Kalman filter methods were introduced as a fault isolation and assessment technique for relative engine performance diagnostics in the late 1970s and early 1980s [76, 77]. The success enjoyed in these early programs encouraged the use of these techniques in subsequent years. Kalman filters have become a popular approach in many current engine performance analysis programs. Some type of Kalman filter is used in a large number of engine diagnosis software, such as Auto Analysis, MAP III (Pratt & Whitney customer test cell data), TEAM III (Pratt & Whitney customer flight data), STORM (Pratt & Whitney self-tuning onboard real-time model), TEMPER (GE customer test cell data), GEM (GE customer flight data), COMPASS (Rolls Royce customer flight data), and GPA (Hamilton standard analysis program). A detailed discussion of the Kalman filter is available from [78–82]. We will briefly review the basic mathematics of the Kalman filter, already discussed in Chapter 1, to facilitate the application to single-fault and sensor fault isolation.

The typical approach followed in engine fault diagnostics involves the use of a linearized model approximation evaluated at a selected engine operating point. This provides a matrix relationship between changes in engine component performance (independent parameters) and the resulting changes in typically measured engine parameters, such as spool speeds, internal temperatures and pressures, fuel flow, etc. (dependent parameters). This relationship can be compactly represented as

$$z = Hx + v \tag{8.1}$$

where z is a vector of measured parameter deltas, x is a vector of fault deltas, H is a matrix of fault influence coefficient relationships between changes in engine component performance, and v is a random vector representing the uncertainties inherent in the measurement process. In addition to the precision of the individual sensors, we also need to address the potential for sensor bias and drift. Therefore, the fault vector given in the model above often relates components directly related to sensor error in addition to engine fault deltas.

The fault vector x given in the model can be partitioned into an engine fault vector (x_e) and a sensor (error) fault vector x_s, i.e., $x = [x_e \vdots x_s]^T$, where

$$x_e = \begin{bmatrix} \Delta\eta_{FAN} \\ \Delta\Gamma_{FAN} \\ \Delta\eta_{CH} \\ \vdots \\ \Delta A_5 \end{bmatrix} \quad \text{and} \quad x_s = \begin{bmatrix} \Delta N1_{err} \\ \Delta P3_{err} \\ \Delta Wf_{err} \\ \vdots \\ \Delta T49_{err} \end{bmatrix} \tag{8.2}$$

We may rewrite Equation (8.1) as

$$z = H_e x_e + H_s x_s + v = \begin{bmatrix} H_e \vdots H_s \end{bmatrix} \begin{bmatrix} x_e \\ \cdots \\ x_s \end{bmatrix} + v = Hx + v \tag{8.3}$$

We see that the matrix H is partitioned into two parts: a matrix of *engine* fault influence coefficients (H_e) and a matrix of *sensor* fault influence coefficients (H_s). The generation of these matrices and their interpretation are discussed in great detail in [9, 83]. The generation of the influence coefficients is done using gas path performance modules. Since this book focuses on signal processing and fault isolation, it is assumed that the influence coefficients are available for a given engine based on thermodynamic models. The model, as given in Equation (8.3), has often been used to track slowly occurring changes in engine performance from revenue flight data, by the use of a Kalman filter-based methodology. Revenue flight data are typical commercial airline flight data; an estimate of these performance shifts, $\hat{x}$, is given by

$$
\begin{bmatrix} \hat{x}_e \\ \cdots \\ \hat{x}_s \end{bmatrix} = \hat{x} = \bar{x} + D(z - H\bar{x}) = \begin{bmatrix} \bar{x}_e \\ \cdots \\ \bar{x}_s \end{bmatrix} + \begin{bmatrix} D_e \\ \cdots \\ D_s \end{bmatrix} \left(z - \left[H_e \vdots H_s \right] \begin{bmatrix} \bar{x}_e \\ \cdots \\ \bar{x}_s \end{bmatrix} \right)
$$

$$
\hat{x} = \text{prediction} + \text{GAIN (residual)} = \text{prediction} + \text{correction} \qquad (8.4)
$$

where $\bar{x}$ represents an *a priori* estimate of the engine/sensor fault deltas and D is the (Kalman) gain matrix referred to as the diagnostic matrix. The diagnostic matrix is computed as a function of several quantities: the engine/sensor influence coefficients H, the measurement covariance matrix R, and a positive semidefinite weighting matrix P_0. The diagnostic matrix is computed as

$$
D = \begin{bmatrix} D_e \\ \cdots \\ D_s \end{bmatrix} = P_0 H^T \left(H P_0 H^T + R \right)^{-1} = \begin{bmatrix} P_0^e H_e^T \left(H_e P_0^e H_e^T + R \right)^{-1} \\ P_0^s H_s^T \left(H_s P_0^s H_s^T + R \right)^{-1} \end{bmatrix} \qquad (8.5)
$$

where P_0^e and P_0^s are weighting submatrices for the engine and sensor fault estimation, respectively. A detailed discussion on the generation of the P_0 and R matrices can be found in reference [9]. Some information was also given in Chapter 1.

The use of the predictor/corrector methods like the Kalman filter to estimate sensor error allows a more reliable and consistent gas turbine module performance analysis. The Kalman filter approach can be applied in a snapshot analysis or as a continuing recursive analysis as new engine data are made available over time. In either case, the simultaneous determination of both engine faults and measurement errors by this methodology has been successfully applied to a large number of commercial and military families of engines with varying instrumentation suites for three decades.

In this chapter, we consider the problem of fault isolation, given the premise that a fault event has been detected. The problem of detection becomes one of recognizing a step or rate change in a gas path parameter or a collection of parameters. The problems associated with fault detection and the mechanisms that can be applied to accomplish this task have been reported in [6]. We have also discussed the trend shift problem in detail in Chapter 6. The types of faults considered in this chapter include engine performance faults, engine system faults, and instrumentation faults.

8.2 Single-Fault Isolation

The Kalman filter can be tailored to behave as a single-fault isolator (SFI). This is a snapshot type of analysis, since it operates on a set of measurement deltas without any *a priori* information or prehistory. The analysis seeks to identify a root cause on the basis of a single-measurement delta set. Root causes are single-fault events and are predefined for the system. They consist of coupled faults within the major modules of the engine, certain system faults such as handling and environmental control system (ECS) bleed leaks and failures, variable stator vane malfunctions, TCC malfunctions, and certain instrumentation faults. A single fault is assumed to occur in isolation. At any given time, there will be one and only one root cause occurrence. There is a finite but low probability of two or more faults occurring at the same time. However, we assume that a typical multiple-fault isolator discussed in Chapter 1 is also processing the data and could be used for such an event. Moreover, algorithmic development needs such assumptions to keep the problem manageable. The aim of the single-fault isolator is to identify the correct root cause once a trend shift is detected. Root causes can be thought of as state variables, $x_1, x_2, \cdots x_n$. They are represented within this system as vectors of measurement deltas, $z_1^*, z_2^*, \cdots, z_n^*$, which are calculated by applying influence coefficients. As an illustration, consider the following 12 root causes.

1	FAN	Coupled FAN (–1% η, –1.25% FC)
2	LPC	Coupled FAN (–1% η, –1.10% FC)
3	HPC	Coupled FAN (–1% η, –0.80% FC)
4	HPT	Coupled FAN (–1% η, +0.75% FP4)
5	LPT	Coupled FAN (–1% η, +1.65% FP45)
6	2.5 BLD	Stability bleed leak (1%)
7	2.9 BLD	Start bleed (1%)
8	FP14	Fan discharge area (1%)
9	FP8	Core discharge area (1%)

10	TCC	Turbine case cooling (on)
11	HPCSVM	HPC stator vane misrigging
12	P49Error	P49 indication problem (2%)

These root causes can exist at varying levels. The state representation of the root cause will be in the form of a 1% cause to be consistent with other influences. An example of the influence coefficients for these root causes is depicted in Table 8.1.

The actual root cause may appear as some multiple of the influences represented in Table 8.1. For instance, a 2.5 bleed root cause may manifest itself as a stuck open bleed ($\approx$15% 2.5 bleed fault) or a partial bleed leak, say, 2% 2.5 bleed faults. These two faults will be treated the same by the Kalman estimator. The difference between the two faults is one of magnitude. The magnitude is estimated by the Kalman filter. However, the ability to estimate correctly depends on the signal-to-noise ratio for the given fault. In this particular case, a 2% 2.5 bleed may get confused with an LPC fault or a 2.9 bleed fault. On the other hand, a stuck 2.5 bleed (14.77%) has a significantly higher SNR and is estimated correctly. Another key factor that governs the accuracy of the estimator is the number of measured parameters. We will consider systems that have between four and eight measurements during the flight, which is typical of most engines.

Given the definitions above for root cause influences, a typical set of possible single faults may be constructed. Consider the following single-fault definitions depicted in Table 8.2. If we represent these 11 faults as $x_1, x_2, \cdots x_{11}$

TABLE 8.1

Sample Root Cause Influences

	T49C2 °C	WF Percent	N2C2 Percent	N1C2 Percent	P25Q2 Percent	T25C2 °C	T3C2 °C	P3Q2 Percent
FAN	3.86	0.7	0.3	-0.68	-2	-1.95	-1.58	-0.03
LPC	-4.54	-0.66	-0.29	-0.14	1.18	0.11	-2.62	0.01
HPC	-6.8	-0.8	0.06	-0.05	-0.83	-0.71	-3.66	0.17
HPT	-10.88	-1.29	0.57	-0.08	-1.29	-1.14	4.03	1.26
LPT	-1.19	0.96	-0.63	0.98	3.4	3.45	-1.42	0.11
2.5 BLD	-3.07	-0.49	-0.16	0	1.04	0.85	-0.86	0
FP14	1.22	0.21	0.07	-0.24	-0.67	-0.73	0.15	-0.01
FP8	-0.61	-1.39	-0.17	-0.64	-1.06	-1.31	-2.62	-1.09
2.9 BLD	-4.22	-1.06	-0.29	-0.06	0.68	0.63	0.6	0.02
TCC	17.75	2.1	-0.9	0.12	2.14	1.86	-4.38	-1.11
HPSCVM	-0.95	-0.11	0.39	0	-0.08	-0.09	0.34	0.09
P49 error	-0.33	-1.7	-0.25	-0.46	-0.55	0.63	0.21	-1.22

Source: Volponi, A.J., et al., *ASME Journal of Engineering for Gas Turbine and Power* 125(4):917–924, 2000. With permission.

TABLE 8.2

Single-Fault Problem Set

Fault	Percent
FAN	−2
LPC	−2
HPC	−2
HPT	−2
LPT	−2
2.5 BLD (low and high)	2 and 14.77
2.9 BLD (low and high)	6.74 and 15.45
HPCSVM	−6
P49 error	2

Source: Volponi, A.J., et al., *ASME Journal of Engineering for Gas Turbine and Power* 125(4):917–924, 2000. With permission.

and the matrix of influences by H^*, an 11×8 matrix, then a set of expected measurement delta vectors, $z_1^*, z_2^*, \cdots, z_{11}^*$, would be calculated by

$$Z_i^* = \left(H^*\right)^T x_i \quad \text{for} \quad i = 1, 2, \cdots 11 \tag{8.6}$$

Each Z_i^* would be an 8×1 vector of measurement deltas.

Recall from Chapter 1 that the general form of the discrete Kalman filter estimator that is used in multiple-fault analysis is as follows, at time $k + 1$:

State extrapolation:

$$\hat{x}(k+1|k) = \Phi(k+1)\hat{x}_k$$

Covariance extrapolation:

$$P(k+1|k) = \Phi(k+1)P_k\Phi^T(k+1) + Q_{k+1}$$

Kalman gain:

$$D_{k+1} = P(k+1|k)H_{k+1}^T\left(H_{k+1}P(k+1|k)H_{k+1}^T + R_{k+1}\right)^{-1}$$

State update:

$$\hat{x}_{k+1} = \hat{x}(k+1|k) + D_{k+1}\left[z_{k+1} - H_{k+1}\,\hat{x}(k+1|k)\right]$$

Covariance update:

$$P_{k+1} = \left[I - D_{k+1}H_{k+1}\right]P[k+1|k]$$

For the single-fault isolator we make the following adjustments:

$$\Phi_{k+1} = I$$

$$Q_{k+1} = 0$$

$$H_{k+1} = M_c \left(H^*\right)^T$$

M_c = measurement configuration vector = diag(m_1, m_2, ..., m_8). Thus, M_c is a diagonal matrix assuming eight potential measurements, where

$$m_j = \begin{cases} 1 & \text{if jth measurement is available} \\ 0 & \text{otherwise} \end{cases}$$

The effect of the measurement configuration matrix is to zero the rows of the root cause influence matrix corresponding to the measurements that are *not* available. The resulting matrix of influences is 8 × 9 for this example.

The single-fault isolation is obtained by processing the general Kalman equations iteratively to provide a snapshot analysis for each of the root causes under consideration (11 in this example). Each "call" to the Kalman filter will be made with a different P_0 matrix chosen to accentuate the kth root cause. Since these are snapshot analyses, the covariance update calculation is not required. The *a priori* state estimate is also assumed to be zero. An SFI analysis is performed typically after a trend shift has been detected in the measurement deltas at some discrete time, say, between time k and k + 1. The delta-delta ($Z_{k+1} - Z_k$) constitutes the input measurement delta, Z, to the SFI.

The SFI is evaluated iteratively for each single fault (11 in the above example). This process yields estimates for each single fault under consideration. It is necessary to rank each of these estimates and determine the top two or three single faults. The measure used to compare estimates for this purpose of ranking is a normalized measurement error norm, which is described below. The single fault admitting the minimum error is deemed the most likely, the second smallest error the next likely, and so forth.

As mentioned above, the normalized measurement error will take into consideration the measurement nonrepeatability of the system. It is assumed that these are known *a priori* or computed from data during initialization of the diagnostic system. However the values are obtained, they are assumed to be known and are passed to the Kalman filter in the form of a positive definite matrix R containing the variances of the measurement deltas (corrected quantities). Thus, if we represent the measurement delta-delta vector between time k and k + 1 assuming that a trend shift has been detected during this time period by $z = [z_1, z_2, \cdots, z_8]$, then the diag$(R) = \left[\sigma_1^2, \sigma_2^2, \cdots, \sigma_8^2\right]$ represents the individual variances. For definiteness, assume we have calculated the nth single-fault estimate. The associated normalized measurement error norm, e_n, is calculated as follows:

$$\hat{x}_n = \text{SFI estimate for the } n\text{th root cause}$$

$$e_n = \left[\frac{\sum\limits_{k=1}^{8} \left(\dfrac{z_k - z_k^*}{\sigma^2} \right)^2}{\sum\limits_{k=1}^{8} z_k^2} \right]^{1/2} \tag{8.7}$$

$$z_k^* = k\text{th element of the vector } M_c \left(H^* \right)^T \hat{x}_n$$

Sometimes, the estimated value for some single faults may be *opposite* in polarity than what would be reasonably expected. For example, an HPT single fault of magnitude –2% might yield an estimated value of +0.5% for an LPC SFI. Since a sudden shift in observed gas path parameters has taken place during engine operation, it is unlikely that the condition of any given module has improved. Therefore, a positive shift in performance does not merit serious consideration. Thus, some preprocessing is necessary before the error is ranked for the fault isolation process. Each of the SFI estimates is analyzed, and only those faults that show a reasonable polarity are shortlisted. Ordering the errors from minimum to maximum yields

$$e_{i_1} \le e_{i_2} \le \cdots \le e_{i_n} \tag{8.8}$$

corresponding to the single-fault estimates

$$\hat{x}_{i_1} \le \hat{x}_{i_2} \le \cdots \le \hat{x}_{i_n} \tag{8.9}$$

in order of likelihood. Typically, the first ranked single fault is selected as the underlying fault and is reported to the user. In some cases, it is better to report the first and second single faults, since erroneous fault identification is possible for situations where the associated error norms are close.

In some situations, the SFI experiences confusion between two single faults. The aliasing of single faults depends on the faults themselves, their relative magnitude with respect to the noise in the measurements, and the number of measurements available for the analysis. All of these factors can contribute to a confounding between two single faults. Some preprocessing of the SFI results may be required to reduce the possibility of false alarms. The rules to be applied are empirically motivated and would be suggested by computer simulation test cases.

8.3 Numerical Simulations

A set of noisy measurements for each of the 11 single faults under consideration (x_i = 1, 2, ..., 11) is generated using Equation (8.1). These noisy measurement vectors (z_k) are then used in the SFI process. Then, the results are ranked and tabulated for first, second, and third ranked faults. These results are shown in Tables 8.3 and 8.4 for eight- and four-measurement set systems, respectively, where the four bleed leak faults have been combined into two faults.

TABLE 8.3

Single-Fault Problem Set Results for Eight Measurements

Fault	1st	Top 2	Top 3
FAN	100%	100%	100%
LPC	90%	100%	100%
HPC	100%	100%	100%
HPT	100%	100%	100%
LPT	100%	100%	100%
2.5 BLD	85.00%	100%	100%
2.9 BLD	96.70%	100%	100%
HPCSVM	100%	100%	100%
P49err	100%	100%	100%
Total	96.90%	100%	100%

Source: Volponi, A.J., et al., *ASME Journal of Engineering for Gas Turbine and Power* 125(4):917–924, 2000. With permission.

TABLE 8.4

Single-Fault Problem Set Results for Four Measurements

Fault	1st	Top 2	Top 3
FAN	100%	100%	100%
LPC	90%	90%	100%
HPC	100%	100%	100%
HPT	100%	100%	100%
LPT	100%	100%	100%
2.5 BLD	50.00%	75%	95%
2.9 BLD	80%	100%	100%
HPCSVM	100%	100%	100%
P49err	100%	100%	100%
Total	91.10%	96.10%	99.40%

Source: Volponi, A.J., et al., *ASME Journal of Engineering for Gas Turbine and Power* 125(4):917–924, 2000. With permission.

The results in the tables show the percentage of time the implanted fault is correctly determined as the first, second, and third choice. The results demonstrate a 96.9 and 91.1% accuracy for the first choice fault isolation for eight- and four-measurement systems, respectively. If we weaken the accuracy criteria to being correct within the top two choices, the relative precision of the isolation increases to 100 and 96.1%, respectively. In either case, the isolation accuracy is quite good.

We see that 91% success rate is possible with only four measurements. It is then natural to ask the question: What is the need for adding more instrumentation? There are several answers to this question. The first answer is that a multiple-fault performance assessment would not perform as well with fewer measurements. But even for the single-fault case, the "more is better" argument remains valid if we consider the robustness of the system. One measure of robustness would be a hard coupled relationship in the implanted faults beyond that which is assumed in the numerical model (i.e., the influences coefficient H^*). Figures 8.1 and 8.2 show the effect on first-choice accuracy for eight- and four-measurement systems, respectively, when the coupling factors for the compression modules FAN, LPC, and HPC are randomly varied [9]. The LPC and HPC clearly exhibit greater robustness to modeling assumptions in the eight-measurement system.

In the next chapter, we will briefly discuss artificial neural networks (ANNs) and their usage in engine performance diagnostics. For the purpose of making a direct comparison with the Kalman filter methodology, we will consider the single-fault isolation problem. The identical computer simulation test will be used as the vehicle for the evaluation.

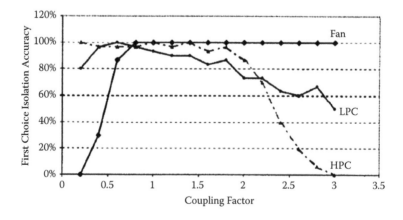

FIGURE 8.1
Coupling factor impact on eight-measurement system. (From Volponi, A.J., et al., *ASME Journal of Engineering for Gas Turbine and Power* 125(4):917–924, 2000. With permission.)

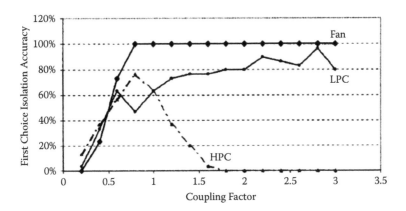

FIGURE 8.2
Coupling factor impact on four-measurement system. (From Volponi, A.J., et al., *ASME Journal of Engineering for Gas Turbine and Power* 125(4):917–924, 2000. With permission.)

8.4 Sensor Error Compensation

We have seen two clear applications of the Kalman filter in gas turbine diagnostics. Broadly, these can be called the multiple-fault isolators (MFIs) and the single-fault isolator (SFI). Recall the measurement model for the MFI:

$$z = H_e x_e + H_s x_s + v \tag{8.10}$$

This equation gives the engine/sensor model at some given engine operating point. Typically, raw engine data that are obtained during flight or during ground tests are normalized to standard day conditions. Next, corrections are applied to each parameter for nonnormal conditions and effects such as engine service bleed, Reynolds and Mach number effects, test cell geometry, and nozzle area deviations [83]. The corrected values are then compared to a reference set of baseline values and the measurement delta is calculated. The above equation is an undetermined system, as the number of unknowns is more than the number of equations. These are m equations due to m measurements, but for each of these measurements, there is an associated sensor fault state. However, since z contains corrected quantities, it leads to a functional dependence on the engine inlet parameters used to normalize them and the base power parameter, e.g., $\Delta WF = f(T_2, P_2, EPR)$. Here T_2, P_2, and EPR are the engine inlet total temperature, pressure, and overall engine pressure ratio, respectively. Since these three parameters also contain errors, the total number of sensor fault deltas is the number of measured parameters plus 3 or $m + 3$. So in total, the number of unknowns for the diagnostic problem is $n_e + m + 3$, where n_e is the number of engine states.

In the Kalman filter MFI, we have m measurements and $n_e + m + 3$ states. It is clear that such a problem does not have a unique algebraic solution. However, an estimate of the unknown states can be obtained by the predictor-corrector Kalman filter. We should point out that the Kalman filter requires a reasonably good answer as a guess. However, this is difficult to get for the engine state, though some empirical estimates have been developed based on engine component deterioration as a function of cycles. However, for sensor faults, the problem of getting good initial estimates is extremely difficult, since there are no good models for predicting bias and drift. Due to this, Kalman filters are often tailored to favor the problem of engine fault estimation. However, this type of approach works only when the measurement errors are small. For such Kalman filter designs, any large sensor error can corrupt all parameter estimates and make them unusable. We will discuss sensor compensation in this section following the method described by Volponi [83].

If we assume that only one large measurement error is there, then the situation is salvageable. Typically, a sensor error occurs in one measurement at a given time, and so this assumption is reasonable. From the Kalman filter equations derived earlier, we get

$$\hat{x}_e = \bar{x}_e + D_e \left(z - H_e \bar{x}_e - H_s \bar{x}_s \right)$$

$$\hat{x}_s = \bar{x}_s + D_s \left(z - H_e \bar{x}_e - H_s \bar{x}_s \right) \tag{8.11}$$

Now, the measurement residual is

$$z - H_e \bar{x}_e = \varsigma = \left[\varsigma_1, \varsigma_2, \varsigma_3, \cdots, \varsigma_m \right]$$

$$\bar{x}_e = \left[e_1, e_2, e_3, \cdots, e_{n_e} \right]$$

$$\bar{x}_s = \left[s_1, s_2, s_3, \cdots, s_{n_s} \right]$$

$$\hat{x}_e = \left[\hat{e}_1, \hat{e}_2, \hat{e}_3, \cdots, \hat{e}_{n_e} \right]$$

$$\hat{x}_s = \left[\hat{s}_1, \hat{s}_2, \hat{s}_3, \cdots, \hat{s}_{n_s} \right]$$

and

$$h_{ij} = \left[H_s \right]_{ij} = ij^{th} \text{ element of } H_s$$

$$d_{ij}^e = \left[D_e \right]_{ij} = ij^{th} \text{ element of } D_e$$

$$d_{ij}^s = \left[D_s \right]_{ij} = ij^{th} \text{ element of } D_s$$

Rewriting Equation (8.11) yields

$$\hat{e}_i = e_i + \sum_{j=1}^{m} d_{ij}^e \left(\varsigma_j - \sum_{n=1}^{ns} h_{jn} s_n \right) (1 \le i \le n_e)$$

$$\hat{s}_i = s_i + \sum_{j=1}^{m} d_{ij}^s \left(\varsigma_j - \sum_{n=1}^{ns} h_{jn} s_n \right) (1 \le i \le n_s)$$

(8.12)

Here $\hat{s}_j$ is the estimated measurement error in the jth engine measurement. If $\hat{s}_j$ is small, then $\hat{e}_j$ will be a reasonable estimate of the jth engine fault. However, if $\hat{s}_j$ is large, there can be misassessment of some or all the engine faults. A decision has to be made to determine if the given sensor delta is sufficiently large.

Consider a threshold T_j and then calculate

$$k = \arg\left\{ \max_n \left(\frac{\|\hat{s}_n - s_n\|}{T_n} \right) \right\}$$

(8.13)

Now the kth sensor is the one with the highest exceedance, and therefore s_k is likely to be a sensor error. We need to find the alternate initial value of s_k'.

Define an index set Ω_k to be the nonzero rows of the kth column of the H_s matrix:

$$\Omega_k = \{n \,|\, h_{nk} \ne 0\}$$

(8.14)

Then Equation (8.12) can be written as

$$\hat{e}_i = e_i + \sum_{n \in \Omega_k} d_{in}^{(e)} \left(\varsigma_n - \sum_m h_{nm} s_m \right) + \sum_{n \notin \Omega_k} d_{in}^{(e)} \left(\varsigma_n - \sum_m h_{nm} s_m \right)$$

(8.15)

The above equation can be written in compact form as

$$\hat{e}_i = e_i + a_i s_k + b_i$$

(8.16)

where

$$a_i = -\sum_{n \in \Omega_k} d_{in}^{(e)} h_{nk}$$

$$b_i = \sum_{n \in \Omega_k} d_{in}^{(e)} \left(\varsigma_n - \sum_{m \notin k} h_{nm} s_m \right) + \varphi_i$$

(8.17)

Note that φ_i, a_i, and b_i are independent of s_k, which means that any change to s_k affects all estimates of the engine in a linear manner. In a similar manner, the second part of Equation (8.12) yields

$$\hat{s}_i = s_i + \alpha_i s_k + \beta_i \tag{8.18}$$

where

$$\alpha_i = -\sum_{n \in \Omega_k} d_{in}^{(e)} h_{nk}$$

$$\beta_i = \sum_{n \in \Omega_k} d_{in}^{(e)} \left(\varsigma_n - \sum_{m \notin k} h_{nm} s_m \right) + \Psi_i \tag{8.19}$$

$$\Psi_i = \sum_{n \in \Omega} d_{in}^{(s)} \left(\varsigma_n - \sum_m h_{nm} s_m \right)$$

Now we choose a new initial value that minimizes the engine and the sensor fault deviation from the expected value, i.e., minimize

$$S = \sum_n \left(\hat{s}_n' - s_n' \right)^2 + \sum_m \left(\hat{e}_m' - e_m' \right)^2 \tag{8.20}$$

where
s_k' is to be specified

$s_n' = s_n \quad \forall n \neq k$
$e_m' = e_m \quad \forall m$

Equations (8.17) and (8.19) are satisfied

To get the minima,

$$\frac{dS}{ds_k} = 0$$

$$s_k' = -\frac{\displaystyle\sum_i \alpha_i \beta_i + \sum_l \alpha_l \beta_l}{\displaystyle\sum_i \alpha_i^2 + \sum_l \alpha_l^2} \tag{8.21}$$

However, we have the following relations:

$$\beta_i = s'_i - s_i - \alpha_i s_k$$

$$\Rightarrow \quad \alpha_i \beta_i = \alpha_i \left(s'_i - s_i \right) - \alpha_i^2 s_k$$

$$\Rightarrow \quad \sum_i \alpha_i \beta_i = \sum_i \alpha_i \left(\hat{s}_i - s_i \right) - \sum_i \alpha_i^2 s_k$$

Similarly,

$$\sum_I a_I b_I = \sum_I a_I \left(\hat{e}_I - e_I \right) - \sum_I a_I^2 s_k$$

Now we put these equations in Equation (8.21) and get

$$s'_k = s_k - \frac{\sum_i \alpha_i \left(\hat{s}_i - s_i \right) + \sum_I \alpha_I \left(\hat{e}_I - e_I \right)}{\sum_i \alpha_i^2 + \sum_I \alpha_I^2}$$

or

$$s'_k = s_k + \frac{[DH_s]_k \cdot \left(\hat{x} - \bar{x} \right)}{\| [DH_s]_k \|^2} \tag{8.22}$$

where $[DH_s]_k$ stands for the kth column of the matrix product DH_s. Also, $\cdot$ is the vector dot product and $\| \ \|$ is the Euclidean norm. Thus, s_k can be adjusted to get the initial value of the kth sensor fault.

This recovery, in case of large sensor error, works well where there is engine deterioration and *a priori* estimates reflect that condition. Recovery from large sensor error is an important part of typical gas turbine diagnostics computer programs.

8.5 Summary

In this chapter, we outlined a Kalman filter-based methodology for fault detection and isolation in gas turbines. Test results have suggested that the Kalman filter method is highly accurate for isolating single gas turbine fault symptoms.

The Kalman filter is a linear model-based estimator and is suitable in those cases where a linear model is available and is known to be a reasonably accurate representation of the input-output relationship. Fortunately, this requirement is satisfied in the engine performance diagnostics application. Years of results from standard gas path analysis software have shown that influence coefficients are fairly accurate and robust linear models. The Kalman filter approach utilizes all model information available, *a priori* estimate information, measurement noise information, etc., and can be easily configured to operate with different measurement suites and fault configurations. For example, it can work for both single-fault and multiple-fault isolation systems. Adaptive measures are also available to allow real-time reconfiguration of the Kalman filter to changing measurement noise levels. An approach to compensate for large sensor error is also discussed in this chapter.

9

Neural Network Architecture

In Chapter 8 we focused on the Kalman filter and its methodology for fault isolation in gas turbines. The present chapter presents the neural networks for fault detection and isolation techniques in gas turbines. Also, a comparison is made between the Kalman filter and neural networks for the fault isolation problem. A special type of neural network named the autoassociative neural network (AANN), which is useful for sensor validation, is also discussed.

In the past decade, artificial neural networks (ANNs) have been employed as a pattern recognition device for gas turbine diagnostics. Both Kalman filter and neural network-based methods have enjoyed reasonable success. The single faults under consideration in this chapter include the engine modules, engine system, and instrumentation faults.

9.1 Artificial Neural Network Approach

An artificial neural network (ANN), usually called neural network (NN), is a mathematical model or computational model that is inspired by the structure or functional aspects of biological neural networks. A neural network consists of an interconnected group of artificial neurons, and it processes information using a connectionist approach to computation. In most cases, an ANN is an adaptive system that changes its structure based on external or internal information that flows through the network during the learning phase. Modern neural networks are nonlinear curve-fitting and statistical data modeling tools. They are often used to model complex relationships between inputs and outputs or to find patterns in data.

The fault isolation problem can be considered to be a pattern classification problem. In such a problem, N-dimensional vectors in an N-dimensional space represent the system response. The system response for different faults tends to be partitioned into different regions of this space and can be regarded as patterns. Pattern recognition involves learning these partitions from simulated or real data, so that a given system response can be classified as a particular fault. Neural networks represent a powerful pattern recognition technique, and have been applied for fault detection of complex systems such as aerospace vehicles [84], nuclear power plants [85], and chemical process facilities [86], among others. A key advantage of neural networks over other methods is their ability to recognize relationships between patterns despite the presence of noise contamination or partial information [87].

Most applications of ANN to fault diagnostics follow a common theme. The ANN is trained offline on fault signatures relating the changes in system measurements from a good baseline to system faults. Typically, faults are embedded into the computer simulations, or real fault data are used, or a combination of both. In case simulated data are used, noise must be added to make the simulations realistic. Such a training process where the ANN is presented input and output data by the system designer is known as supervised learning. Once the ANN has been properly trained using this process of supervised learning, it can analyze data that are different from those it was originally exposed to during the training sessions. When the trained ANN is placed online, it recognizes a similar response from the actual system.

The discussion below involves the use of two types of neural networks for engine fault diagnostics: a feed-forward ANN trained using a back-propagation (BP) algorithm and a hybrid neural network. The use of such networks for gas turbine diagnostics was proposed by Volponi, Depold, Ganguli, and Daguang [12] in 2003, and the present chapter is largely based on this paper.

9.1.1 Back-Propagation (BP) Algorithm

While there are several types of neural networks, multilayer feed-forward networks trained using the back-propagation algorithm have emerged as the most widely used. Figure 9.1 illustrates the schematic of a feed-forward neural network, which consists of an input layer, an output layer, and one or more hidden layers. The number of neurons in the input and output layers is determined by the number of input measurements and output parameters. The number of hidden layer nodes is selected based on a convergence criterion and the characteristic of the input-output mapping relationship.

A three-layer feed-forward network is used for the present chapter, as shown in Figure 9.1. The feed-forward network is trained using supervised learning, which involves presenting input-output pairs to the ANN and then using the BP algorithm to learn the relationship between the inputs and outputs by minimizing the following error measure:

$$E = \sum_{k=1}^{N} E_k \qquad (9.1)$$

in which E_k represents the root mean square error associated with the kth training sample and N represents the number of samples that are used for training the network. The BP algorithm uses a gradient search to perform the nonlinear optimization needed to minimize the error [88].

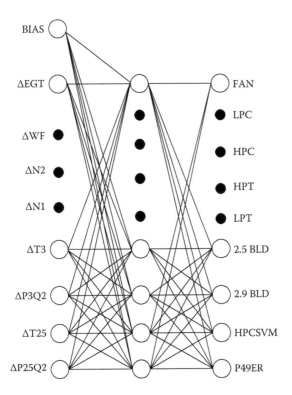

FIGURE 9.1
Architecture of three-layer feed-forward ANN. (From Volponi, A.J., et al., *ASME Journal of Engineering for Gas Turbine and Power* 125(4):917–924, 2000. With permission.)

To improve the ANN's ability to deal with data scatter, the input data were normalized using the following formula:

$$Y_{in} = Y_{im} / (Y_{i\,max} \cdot \sigma_i)$$

where Y_i is the ith monitoring parameter; n, m, and max are the normalized, measured, and maximum possible values, respectively; and σ_i is the standard deviation of the ith monitoring parameter.

The standard deviations and the influence coefficients used for the ANN testing are the same as those used for the Kalman filter described in previous chapters. For comparison of fault isolation results with 4 inputs and 8 inputs, 20 training cases and 50 testing cases were generated. The training cases were used for the BP algorithm to train the neural network. Once the neural network was trained, the test cases were used to evaluate the performance of the neural network. Data used for training were not used for testing the neural network. The diagnostic results were considered for the three highest outputs, which represent the three most likely faults. The test results for eight and four measurements using the BP ANN are shown in Tables 9.1 and 9.2, respectively. For both the eight- and four-measurement cases, the Kalman single-fault isolator (SFI) results shown in Tables 8.3 and 8.4 in Chapter 8 are better than the BP ANN results.

For the eight-measurement case, the Kalman SFI has 100% accuracy in fault isolation among the top three choices, compared to 95.7% for the BP ANN. For the four-measurement case, the Kalman SFI has an accuracy of 99.4% in fault isolation among the top three choices, compared to 94.6% for the BP ANN.

TABLE 9.1

SFI Accuracy: ANN, Eight Measurements

Fault	1st	Top 2	Top 3
FAN	100%	100%	100%
LPC	60%	80%	90%
HPC	100%	100%	100%
HPT	100%	100%	100%
LPT	100%	100%	100%
2.5 BLD	80%	85%	85%
2.9 BLD	86.70%	86.70%	86.70%
HPCSVM	90%	100%	100%
P49err	100%	100%	100%
Total	90.70%	94.60%	95.70%

Source: Volponi, A.J., et al., *ASME Journal of Engineering for Gas Turbine and Power* 125(4):917–924, 2000. With permission.

TABLE 9.2

SFI Accuracy: ANN, Four Measurements

Fault	1st	Top 2	Top 3
FAN	100%	100%	100%
LPC	90%	90%	90%
HPC	90%	100%	100%
HPT	100%	100%	100%
LPT	100%	100%	100%
2.5 BLD	75%	75%	75%
2.9 BLD	86.70%	86.70%	86.70%
HPCSVM	100%	100%	100%
P49err	100%	100%	100%
Total	93.50%	94.60%	94.60%

Source: Volponi, A.J., et al., *ASME Journal of Engineering for Gas Turbine and Power* 125(4):917–924, 2000. With permission.

9.1.2 Hybrid Neural Network Algorithm

The BP ANN is handicapped relative to the Kalman SFI in some ways. For example, while the Kalman SFI uses influence coefficients in the form of a matrix to define the model, the BP ANN uses data generated from influence coefficients to learn the model.

A hybrid NN is a network architecture where one or more ANN functions are replaced by an algorithm that includes domain knowledge [89]. The objective is to substitute features in the neural network architecture that are already analytically understood. This avoids the need for training the ANN to learn information that is already known. For example, instead of using training data based on influence coefficients, the hybrid ANN uses the influence coefficients as part of the network model.

A Gaussian nearest-neighbor function was substituted for the ANN feature identification function, and the fault mapping function was ignored for this hybrid NN algorithm. The output of the network was the root sum of the squares number of standard deviations from a perfect match of the fault pattern. The output of the network was used directly to rank the faults.

A test was made to determine if the neural network mapping to the faults could also be optimized. Up to 72 weightings were available between the eight-measurement features and the nine faults. Thirty-six weightings were available with four measurements. While it was demonstrated that the weightings could be optimized, the tests were run with all the weightings set to unity. The results from the hybrid neural network are shown in Tables 9.3 and 9.4, for eight and four measurements, respectively.

The accuracy of the hybrid neural network compares favorably with the Kalman SFI. For the eight-measurement case, both the Kalman SFI and

TABLE 9.3

SFI Accuracy: Hybrid NN, Eight Measurements

Fault	1st	Top 2	Top 3
FAN	100%	100%	100%
LPC	90%	100%	100%
HPC	100%	100%	100%
HPT	100%	100%	100%
LPT	100%	100%	100%
2.5 BLD	90%	100%	100%
2.9 BLD	100%	100%	100%
HPCSVM	100%	100%	100%
P49err	100%	100%	100%
Total	97.80%	100%	100%

Source: Volponi, A.J., et al., *ASME Journal of Engineering for Gas Turbine and Power* 125(4):917–924, 2000. With permission.

TABLE 9.4

SFI Accuracy: Hybrid NN, Four Measurements

Fault	1st	Top 2	Top 3
FAN	90%	100%	100%
LPC	90%	90%	100%
HPC	100%	100%	100%
HPT	100%	100%	100%
LPT	100%	100%	100%
2.5 BLD	70%	90%	86.70%
2.9 BLD	86.70%	86.70%	100%
HPCSVM	70%	100%	100%
P49err	100%	100%	100%
Total	91.10%	97.80%	98.40%

Source: Volponi, A.J., et al., *ASME Journal of Engineering for Gas Turbine and Power* 125(4):917–924, 2000. With permission.

the hybrid ANN show fault isolation accuracy among the top three choices of 100%. For the four-measurement case, the Kalman SFI shows fault isolation accuracy among the top three choices of 99.4%, compared to 98.4% for the hybrid ANN.

In general, the hybrid ANN gives better results than the BP ANN. This may be because the hybrid uses the influence coefficients for each fault within the network, whereas the BP ANN has to learn the influence coefficients from the training data.

9.2 Kalman Filter and Neural Network Methods

Feed-forward neural networks are typically made of interconnected nonlinear neurons and are particularly useful where the input-output relationship is nonlinear. The Kalman filter and the hybrid neural network assume a linearized model of the system. In the present chapter, the linear model takes the form of influence coefficients. For a feed-forward ANN, BP learning for linear problems in engine fault diagnostics will generally result in poor performance compared to a Kalman filter or hybrid ANN approach.

Learning in neural networks involves mapping an input to an output. The inputs and outputs can be generated from models, from real data, or a combination of both. ANNs can therefore also be model-free estimators, a quality that is very useful if modeling information such as influence coefficients is not available. Kalman filters, on the other hand, are model-based estimators, and are suitable for problems in engine performance diagnostics where influence coefficients are available as the model.

Neural networks are not limited to multilayer neurons with BP training. Self-organizing maps based on competitive learning [90], simulated annealing based on statistical thermodynamics [91], Boltzmann learning [92], and radial basis functions [93] are some of the other developments in ANN that may be applicable to engine diagnostics. Combining the Kalman filter approach with some of the ANN methods may yield superior results for the engine diagnostic problem than those obtainable within either methodology acting alone.

9.3 Autoassociative Neural Network

We saw in the previous chapter that in addition to degradation in the engine components, there also exists the possibility of change or degradation in the sensors.

In case a sensor fails, it would be nice to get a good estimate of the faulty sensor. As discussed in Chapter 8, the Kalman filter logic can be suitably adjusted to get this estimate. The AANN is a special type of neural network that is very suitable for sensor validation. Its use for gas turbine diagnostics was suggested by Guo and Saus [94].

The basic architecture of the AANN is shown in the Figure 9.2. The network consists of an input layer, an output layer, and three hidden layers.

The input and output layers have the same number of sensors, which are equal to the number of gas turbine measurements. The middle layer, which typically has a low number of neurons, is called the bottleneck layer. The measurement data are compressed into the bottleneck layer and then regenerated or expanded thereafter. The number of neurons in the bottleneck layer represents the degrees of freedom of the measurement set, i.e., the minimum number or principal components of the measurements.

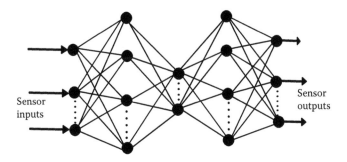

FIGURE 9.2
Autoassociative neural network for sensor validation.

The AANN is trained to learn the relationship between the sensor measurements. The different sensor measurements are typically related to one another by some mathematical relationship. If any one sensor measurement is out of line, the inconsistency will be visible from these equations. The AANN uses the learned relationship between the sensors to recreate an estimate of the incorrect measurement.

Standard algorithms such as the back-propagation algorithm can be used to train the network. The sigmoid activation function can be used, and the training should ensure that the desired sensor measurement is returned for both normal data and training data.

Sensor failure is detected as follows: Each normal network output, say, $\hat{z}_i$, is compared with the corresponding input z_i. Let us call this $\varepsilon_i = |\hat{z}_i - z_i|$. If any single ε_i changes substantially compared to a given threshold for that sensor T_i, but the other error estimates $\varepsilon_j = |\hat{z}_j - z_j|$, $j \neq i$, stay low, then the ith sensor measurement is faulty.

After a faulty sensor is detected, the corresponding measurement is removed from the input layer. The AANN now retains the capacity to synthesize this missing faulty sensor measurement from the other sensor data. This estimated value also comes in handy if it is used for control functions.

We can see that the AANN uses the concept of dimensional reduction. Due to physics, it is generally not possible for the sensor measurements to be completely independent of each other. There is some redundancy between the sensor measurements. However, the process of coming up with the appropriate number of neurons in the three hidden layers is key to the proper functioning of the AANN.

9.4 Summary

Numerical simulations show that the back-propagation neural network, the hybrid neural network, and the Kalman filter method are highly accurate for isolating single gas turbine faults. Furthermore, the results also indicate that these methodologies compare favorably in terms of accuracy, with a very slight advantage going to the Kalman filter approach.

Each method has its own advantages and disadvantages. The ANN is inherently nonlinear and can be used in applications where model information is scarce or lacking altogether. The ANNs are, however, data driven and therefore must be trained. The training is typically performed offline in a supervised fashion, meaning that the input-output relationship is known. This means that the underlying faults in the training data are already known. This could be a drawback if real engine data were used for training, since the precise nature of the fault may or may not be known. If the engine configuration or sensor noise levels change, the ANN approach requires that the ANN be retrained. Once trained, the ANN architecture provides a numerically simple and hence fast diagnostic operator suitable for real-time application.

The hybrid neural network, like the Kalman filter, is a model-based method that uses influence coefficients as the primary linear model in a neural network architecture framework. For the single-fault isolation problem referenced in this chapter, the hybrid NN closely resembles a weighted least-squares solution. This explains the close agreement between the hybrid NN and Kalman filter in the single-fault isolation numerical simulations.

The autoassociative neural network uses the concepts from principal component analysis to learn the relationship between the sensor measurements. It is a useful tool for detection of sensor faults and for getting a reasonable estimate of their value.

10

Fuzzy Logic System

Chapters 8 and 9 introduced the Kalman filter and the neural network for fault isolation in gas turbine engines, respectively. Fuzzy expert systems are robust and are being increasingly used for diagnostics and other applications [95–98]. Recently, it has been proven that classical feed-forward neural networks of the type used in engine diagnostics can be approximated to an arbitrary degree of accuracy by a fuzzy logic system, without having to go through the laborious training process needed by a neural network [99]. In addition, fuzzy rules follow human language-based reasoning processes and are much easier to interpret and understand than neural networks that have a black box nature [100]. In this chapter, it is shown that fuzzy logic systems can be used for engine module fault isolation under high levels of uncertainty. In addition, fault isolation results from the fuzzy logic systems are compared with results from neural networks and Kalman filter methods. The application of fuzzy logic to gas turbine diagnostics, which is discussed in this chapter, was introduced by Ganguli [13, 101].

10.1 Module and System Faults

Most damages to a typical, twin-spool gas turbine engine shown in Figure 1.2 manifest themselves as changes in either the module efficiency or flow capacity/area. The FAN, LPC, and HPC modules have flow capacities associated with them. The HPT and LPT modules have areas associated with them. Besides the five modules, the engine can experience system faults such as bleed leaks and failures, variable stator vane malfunctions, and certain instrumentation faults. The fault models for the nine single faults considered in this chapter are shown in Table 10.1.

Besides the five module faults, other faults considered are the start bleed leak (2.9 BLD), stability bleed leak (2.5 BLD), P49 indication problem (P49ER), and stator vane misrigging (HPCSVM). It is assumed that one and only one single fault occurs at a given time. The fingerprints for each of the nine faults are shown in Table 10.2.

Figures 10.1 and 10.2 show an example of a fingerprint chart for the low-pressure turbine fault and 2.5 bleed fault, respectively. Figures 10.1 and 10.2 are obtained from the fingerprint charts in Table 10.2. From these charts, it appears that it would be difficult for a human engineer to look at several

TABLE 10.1

Description and Modeling of Single Faults

Fault	Description	Model
FAN	Damage in fan module	−2% η, −2.5% FC
LPC	Damage in LPC module	−2% η, −2.2% FC
HPC	Damage in HPC module	−2%η, −1.6% FC
HPT	Damage in HPT module	−2% η, +1.5% FP4
LPT	Damage in LPT module	−2% η, +3.3% FP45
2.5 BLD	Stability bleed leak	2%
2.9 BLD	Start bleed leak	2%
HPCSVM	HPC stator vane misrigging	−6%
P49ER	P49 indication problem	2%

Source: Ganguli, R., *Journal of Propulsion and Power* 18(2):440–447, 2002. With permission.

TABLE 10.2

Fingerprints for Selected Gas Turbine Faults

Measurement Faults	$\Delta EGT,$ ±°C	$\Delta WF,$ %	$\Delta N2,$ %	$\Delta N1,$ %	$\Delta P25,$ %	$\Delta T25,$ ±°C	$\Delta T3,$ ±°C	$\Delta P3,$ %
HPC	13.60	1.6	−0.11	0.1	1.66	1.41	7.31	−0.34
HPT	21.77	2.58	1.13	0.15	2.59	2.28	−8.05	−2.52
LPC	9.09	1.32	0.57	0.28	−2.35	−0.22	5.23	−0.01
LPT	2.38	−1.92	1.27	−1.96	−6.80	−6.90	2.84	−0.22
FAN	−7.72	−1.40	−0.59	1.35	3.99	3.91	3.15	0.05
2.5 bleed	6.15	0.99	0.31	0.01	−2.08	−1.71	1.73	0.01
2.9 bleed	8.43	2.12	0.58	0.12	−1.36	−1.27	−1.20	−0.04
HPCSVM	−5.71	−0.69	2.33	−0.02	−0.51	−0.54	2.05	0.54
P49ER	−0.65	−3.40	−0.49	−0.92	−1.11	0.03	−0.42	−2.44

Source: Ganguli, R., *Journal of Propulsion and Power* 18(2):440–447, 2002. With permission.

such charts and identify the correct fault from the measurement deltas. However, this task is well suited for a fuzzy logic system.

10.2 Fuzzy Logic System

A fuzzy logic system is a nonlinear mapping of an input feature vector into a scalar output [102]. Fuzzy set theory and fuzzy logic provide the framework for the nonlinear mapping. Fuzzy logic systems have been widely used in engineering applications because of the flexibility they offer designers and their ability to handle uncertainty. A fuzzy logic system can be expressed as a linear combination of fuzzy basis functions and is a universal function approximater.

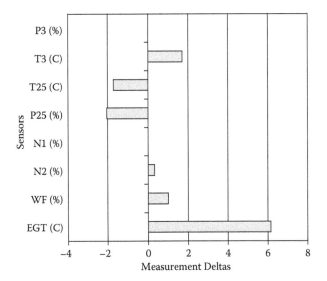

FIGURE 10.1
Fingerprint chart for 2.5 bleed fault. (From Ganguli, R., *Journal of Propulsion and Power* 18(2):440–447, 2002. With permission.)

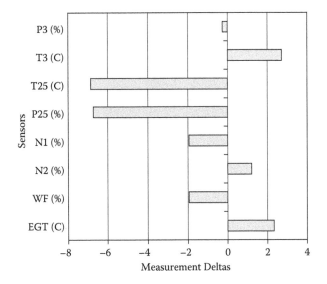

FIGURE 10.2
Fingerprint chart for LPT fault. (From Ganguli, R., *Journal of Propulsion and Power* 18(2):440–447, 2002. With permission.)

A typical multi-input single-output (MISO) fuzzy logic system performs a mapping using four basic components: rules, fuzzifier, inference engine, and defuzzifier. Here

$$f: V \in R^m \rightarrow W \in R \tag{10.1}$$

where

$$V = V_1 \times V_2 \times \cdots \times V_n \in R^m \tag{10.2}$$

is the input space and $V \in R$ is the output space.

A fuzzy logic system maps inputs to outputs using four basic components: rules, fuzzifier, inference engine, and defuzzifier are shown in Figure 10.3. Once the rules governing the fuzzy logic system have been fixed, the fuzzy logic system can be expressed as a mapping of inputs to outputs.

Rules can come from experts or can be obtained from numerical data. In either case, engineering rules are expressed as a collection of IF-THEN statements, such as "IF u_1 is HIGH, and u_2 is LOW, THEN v is LOW." To formulate such a rule we need an understanding of:

1. Linguistic variables vs. numerical values of a variable (e.g., HIGH vs. 3.5%).

2. Quantifying linguistic variables, e.g., may have a finite number of linguistic terms associated with them, ranging from NEGLIGIBLE to VERY HIGH, which is done using fuzzy membership functions.

3. Logical connections between linguistic variables, e.g., AND, OR, etc.

4. Implications such as "IF A, THEN B." It is also necessary to combine more than one rule.

The fuzzifier maps crisp input numbers into fuzzy sets. It activates rules that are expressed in terms of linguistic variables. An inference engine of the fuzzy logic system maps fuzzy sets to fuzzy sets and determines the way in which the fuzzy sets are combined. In several applications, crisp numbers

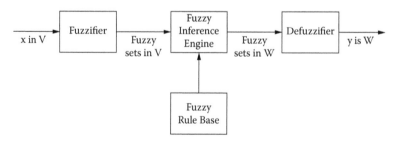

FIGURE 10.3
Schematic representation of a fuzzy logic system. (From Ganguli, R., *Journal of Propulsion and Power* 18(2):440–447, 2002. With permission.)

are needed as an output of the fuzzy logic system. In those cases, a defuzzifier is used to calculate crisp values from fuzzy values.

Fuzzy sets. A fuzzy set F is defined on a universe of discourse U and is characterized by a degree of membership $\mu(x)$, which can take on values between 0 and 1. A fuzzy set generalizes the concept of an ordinary set whose membership function takes only two values, zero and unity.

Linguistic variables. A linguistic variable u is used to represent the numerical value x, where x is an element of U. A linguistic variable is usually partitioned into a set of terms $T(u)$, which cover its universe of discourse.

Membership functions. The most commonly used shapes for membership functions $\mu(x)$ are triangular, trapezoidal, piecewise linear, or Gaussian. The designer selects the type of membership function used. There is no requirement that membership functions overlap. However, one of the major strengths of fuzzy logic is that membership functions can overlap. Fuzzy logic systems are robust because decisions are distributed over more than one input class. For convenience, membership functions are normalized to 1, so they take values between 0 and 1, and thus define the fuzzy set [103].

Inference engine. Rules for the fuzzy system can be expressed as

$$R_i : \text{IF } x_1 \text{ is } F_1 \text{ AND } x_2 \text{ is } F_2 \text{ AND } \cdots \text{AND } x_m \text{ is } F_m$$

$$\text{THEN } y = C_i \quad i = 1,2,3\cdots M \tag{10.3}$$

where m and M are the number of input variables and rules, x_i and y are the input and output variables, and $F_i \in V_i$ and $C_i \in W$ are fuzzy sets characterized by membership functions $\mu_{F_i}(x)$ and $\mu_{C_i}(x)$, respectively. Each rule can be viewed as a fuzzy implication:

$$F_{1,2,\cdots,m} = F_1 \times F_2 \times \cdots \times F_m \rightarrow C_i \tag{10.4}$$

which is a fuzzy set in $V \times W = V_1 \times V_2 \times \cdots \times V_m \times W$ with membership function given by

$$\mu_{R_i}(x,y) = \mu_{F_1}(x_1) * \mu_{F_2}(x_2) * \cdots * \mu_{F_m}(x_m) * \mu_{C_i}(y) \tag{10.5}$$

where $*$ is the T-norm with $x = [x_1, x_2, \cdots, x_m] \in V$ and $y \in W$. This sort of rule is suitable for many applications. The algebraic product is one of the most widely used T-norms in applications, and leads to a product inference engine. In pattern recognition problems, the outputs are often crisp sets and $\mu_{C_i}(y) = 1$ is often used for the product inference formula [102].

10.3 Defuzzification

Popular defuzzification methods include maximum matching and centroid defuzzification. Centroid defuzzification is widely used for fuzzy control problems where a crisp output is needed. Maximum matching is often used for pattern matching problems where we need to know the output class. Suppose there are K fuzzy rules and among them K_j rules ($j = 1, 2, \ldots, L$, where L is the number of classes) produce class C_j. Let D_p^i be the measurements of how the pth pattern matched the antecedent conditions (IF part) of the ith rule, which is given by the product of membership grades of the pattern in the regions that the ith rule occupies:

$$D_p^i = \prod_{i=1}^{m} \mu_{li} \tag{10.6}$$

where m is the number of inputs and μ_{l_i} is the degree of membership of measurement l in the fuzzy regions that the ith rule occupies. Let $D_p^{\max}\left(C_j\right)$ be the maximum matching degree of the rules (rules j_l, $l = 1, 2, \ldots, K_j$) generating class C_j:

$$D_p^{\max}\left(C_j\right) = \max_{l=1}^{K_j} D_p^{j_l} \tag{10.7}$$

Then the system will output class C_j^* provided that

$$D_p^{\max}\left(C_j^*\right) = \max_j D_p^{\max}\left(C_j\right) \tag{10.8}$$

If there are two or more classes that achieve the maximum matching degree, we will select the class that has the largest number of fired fuzzy rules (a fired rule has a matching degree of greater than zero).

10.4 Problem Formulation

10.4.1 Input and Output

Inputs to the fuzzy logic systems are measurement deltas and outputs are engine module faults. We have eight measurements represented by z and nine engine faults represented by ξ. The objective is to find a functional mapping between z and ξ. Mathematically, this can represented as

$$\xi = F(z) \tag{10.9}$$

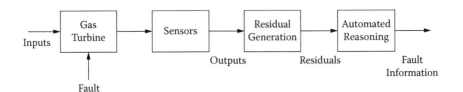

FIGURE 10.4
Schematic representation of intelligent fault isolation system for gas turbine. (From Ganguli, R., *Journal of Propulsion and Power* 18(2):440–447, 2002. With permission.)

where

$$\xi = \{\text{FAN, LPC, HPC, HPT, LPT, 2.5 BLD, 2.9 BLD, HPCSVM, P49ER}\}^T$$

and

$$z = \{\Delta EGT, \Delta WF, \Delta N2, \Delta N1, \Delta P25, \Delta T25, \Delta P3, \Delta T3\}^T \qquad (10.10)$$

Each measurement delta has uncertainty coming from both modeling errors and measurement errors, which makes the preceding inverse problem difficult to solve.

A schematic representation of the intelligent system for gas turbine fault isolation is shown in Figure 10.4. The gas turbine in its normal state of operation can be viewed as an input-output system operating in steady state. When a single fault occurs, there is a sharp change in the gas path sensor measurements reflecting the change in the gas turbine. This change in sensor measurement can then be compared to the baseline engine without faults to obtain a measurement residual. The residuals generated by the faulty system then are subjected to automated reasoning by fuzzy logic to yield fault information.

10.5 Fuzzification

Here FAN, LPC, HPC, HPT, and LPT are fuzzy sets denoting the five engine modules. In addition, 2.5 BLD, 2.9 BLD, HPCSVM, and P49ER are fuzzy sets denoting the four system faults. Each fuzzy set has degrees of membership ranging from 0 to 1. In this chapter, we are only interested in isolating the fault and in its magnitude. Therefore, we do not further decompose the module fuzzy sets using linguistic variables.

The measurement deltas ΔEGT, $\Delta N1$, $\Delta N2$, ΔWF, $\Delta P25$, $\Delta T25$, $\Delta P3$, and $\Delta T3$ are also treated as fuzzy variables. To get a high degree of resolution, they are further split into linguistic variables. For example, consider ΔEGT as a linguistic variable. It can be decomposed into a set of terms:

$$T(\Delta EGT) = \{\text{very high--, high--, medium high--, medium--, medium low--,}$$
$$\text{low medium--, low--, negligible, low+, low medium+, medium}$$
$$\text{low+, medium+, medium high+, high+, very high+}\} \qquad (10.11)$$

where each term in $T(\Delta EGT)$ is characterized by a fuzzy set in the universe of discourse $U(\Delta EGT) = \{-25°C, 25°C\}$, which is selected to include values spanning the fingerprint charts in Table 10.2, while maintaining symmetry. A total of 15 fuzzy sets are used to partition the numerical variables. It is found that a courser partition does not give very accurate results. A trial and error process and a careful study of the fingerprint charts were used to obtain the number of fuzzy sets. This process can be labeled heuristic reasoning.

The other seven measurement deltas are defined using the same set of terms as ΔEGT, spanning the following universes of discourse:

$$U(\Delta WF) = \{-4, 4\%\}; \ U(\Delta N2) = \{-3, 3\%\}; \ U(\Delta N1) = \{-3, 3\%\};$$

$$U(\Delta P25) = \{-6, 6\%\}; \ U(\Delta T25) = \{-10, 10°C\}; \qquad (10.12)$$

$$U(\Delta T3) = \{-10, 10°C\}; \ U(\Delta P3) = \{-3, 3\%\}$$

Since the influence coefficients on which the fingerprints are based represent a linear model, the diagnostic system should be limited to small measurement deltas. In addition, measurement deltas larger than covered by the universe of discourse will represent a large fault indicative of a catastrophic failure.

Fuzzy sets with Gaussian membership functions are used. These fuzzy sets can be defined using the following equation:

$$\mu(x) = e^{-0.5((x-m)/\sigma)^2} \qquad (10.13)$$

where m is the midpoint of the fuzzy set and σ is the uncertainty (standard deviation) associated with the variable.

Table 10.3 gives the linguistic measure associated with each fuzzy set and the midpoint of the set for each measurement delta.

The midpoints are selected to span the region ranging from a perfect engine (all measurement deltas are zero) to one with significant damage.

$$\mu(x) = e^{-0.5(x-m/\sigma)^2} \qquad m_{VH-} < x \ \text{ OR } \ x < m_{VH+}$$

$$\mu(x) = 1 \qquad\qquad m_{VH+} < x \ \text{ OR } \ x < m_{VH-}$$

$$(10.14)$$

TABLE 10.3

Midpoints of Gaussian Fuzzy Sets

Linguistic Measure	Symbol	Measurement Deltas					
		$\Delta EGT,$ ±°C	$\Delta WF,$ %	$\Delta N1$ and $\Delta N2$, %	$\Delta P25,$ %	$\Delta T25$ and $\Delta T3$, °C	$\Delta P3,$ %
Very high–	VH–	–20	–3	–2	–6	–10	–3
High–	H–	–15	–2.5	–1.5	–4.5	–7.5	–2.5
Medium high–	MH–	–12.5	–2.25	–1.25	–3	–5	–2.25
Medium–	M+	–10	–2	–1	–2.5	–2.5	–1.5
Medium low–	ML–	–7.5	–1.5	–0.5	–2	–1.75	–0.75
Low medium–	LM–	–5	–1	–0.25	–1.5	–1.25	–0.5
Low–	L–	–2.5	–0.75	–0.125	–0.5	–0.25	–0.25
Negligible	N	0	0	0	0	0	0
Low+	L+	2.5	0.75	0.125	0.5	0.25	0.25
Low medium+	LM+	5	1	0.25	1.5	1.25	0.5
Medium low+	ML+	7.5	1.5	0.5	2	1.75	0.75
Medium+	M+	10	2	1	2.5	2.5	1.5
High+	H+	12.5	2.25	1.25	3	5	2.25
Medium high+	MH+	15	2.5	1.5	4.5	7.5	2.5
Very high+	VH+	20	3	2	6	10	3

Source: Ganguli, R., *Journal of Propulsion and Power* 18(2):440–447, 2002. With permission.

TABLE 10.4

Measurement Uncertainty

Measurement Delta	Standard Deviation
ΔEGT	4.23°C
$\Delta N1$	0.25%
$\Delta N2$	0.17%
ΔWF	0.50%
$\Delta P25$	0.46%
$\Delta T25$	1.12°C
$\Delta T3$	1.99°C
$\Delta P3$	0.24%

Source: Data from Ganguli, R., *Journal of Propulsion and Power* 18(2):440–447, 2002.

Here m_{VH+} represents the midpoint corresponding to fuzzy set VH+. The standard deviations for the measurement deltas are representative of airline data [11] and are shown in Table 10.4. For illustration, Figure 10.5 shows the membership functions for each of the 15 fuzzy sets for $\Delta N2$.

The membership functions for the other fuzzy sets are similar in appearance. The midpoints for these fuzzy sets are obtained through heuristic reasoning.

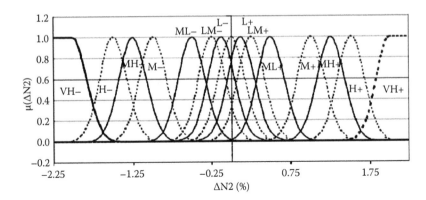

FIGURE 10.5

Fuzzy sets for high rotor speed (*N2*) measurement delta. (From Ganguli, R., *Journal of Propulsion and Power* 18(2):440–447, 2002. With permission.)

10.6 Rules and Fault Isolation

Rules for the fuzzy system are obtained by fuzzification of the numerical values in the fingerprint charts using the following procedure [104, 105]:

1. A set of eight measurement deltas corresponding to a given module fault is input to the fuzzy logic system (FLS) and the degree of membership of the elements of ΔEGT, ΔWF, $\Delta N2$, $\Delta N1$, $\Delta P25$, $\Delta T25$, $\Delta T3$, and $\Delta P3$ is obtained. Therefore, each measurement has 15 degrees of memberships based on the linguistic measures in Table 10.3.

2. Each measurement delta is then assigned to the fuzzy set with the maximum degree of membership.

3. One rule is obtained for each module fault by relating the measurement delta with the maximum degree of membership to a module fault.

These rules are tabulated in Table 10.5. The linguistic symbols used in this table are defined in Table 10.3. These rules can be read as follows for the FAN module:

IF

ΔEGT is medium low– AND

ΔWF is medium low– AND

$\Delta N2$ is medium low– AND

$\Delta N1$ is medium high+ AND

TABLE 10.5

Rules for Fuzzy System

Measurement Faults	ΔEGT	ΔWF	$\Delta N2$	$\Delta N1$	$\Delta P25$	$\Delta T25$	$\Delta T3$	$\Delta P3$
HPC	MH+	ML+	L–	L+	LM+	LM+	H+	L–
HPT	VH+	H+	MH–	L+	M+	M+	H–	H–
LPC	M+	ML+	ML+	LM+	M–	L–	MH+	N
LPT	L+	M–	MH+	VH–	VH–	H–	M+	L–
FAN	ML–	ML–	ML–	MH+	H+	MH+	M+	N
2.5 bleed	LM+	LM+	LM+	N	ML–	ML–	ML+	N
2.9 bleed	ML+	MH+	ML+	L+	LM–	LM–	LM–	N
HPCSVM	LM–	L–	VH+	N	L–	L–	ML+	LM+
P49ER	N	VH–	ML–	M–	LM–	N	L–	H–

Source: Ganguli, R., *Journal of Propulsion and Power* 18(2):440–447, 2002. With permission.

> $\Delta P25$ is high+ AND
> $\Delta T25$ is medium high+ AND
> $\Delta T3$ is medium+ AND
> $\Delta P3$ is negligible
> THEN
> Problem in FAN module

The rules for the other faults can be similarly interpreted. These rules provide a knowledge base and represent how a human engineer would interpret data to isolate an engine fault using fingerprint charts.

The fuzzy rules in Table 10.5 represent a fuzzified model of the fingerprints shown in Table 10.2. Because Gaussian fuzzy sets asymptotically approaching zero far from the midpoint are used, all of the rules fire at some level. For any given input set of measurement deltas, the fuzzy rules are applied using product implication. Once the fuzzy rules are applied for a given measurement, we have degrees of membership for each of the nine faults. For fault isolation, we are interested in the most likely fault. The fault with the highest degree of membership is selected as the most likely fault.

10.7 Numerical Simulations

The fuzzy system is tested using simulated data developed from the fingerprint charts. For each fault, 1000 data sets are generated. Noise is added to the simulated measurement deltas using the typical standard deviations

TABLE 10.6

Fault Isolation Results (%) from Fuzzy System

Measurement Faults	Basic Four	Basic Four +P25 +T25	Basic Four +P3 +T3	All Eight
HPC	92	100	98	100
HPT	100	100	100	100
LPC	82	85	94	94
LPT	100	100	100	100
FAN	100	100	100	100
2.5 bleed	69	78	82	89
2.9 bleed	61	93	94	99
HPCSVM	100	100	100	100
P49ER	100	100	100	100
Average success rate	89	95	96	98

Source: Ganguli, R., *Journal of Propulsion and Power* 18(2):440–447, 2002. With permission.

shown in Table 10.4. Testing with noisy data allows an analysis of the robustness of the system.

Table 10.6 shows the test results from the fuzzy system. The accuracy of fault detection is shown for the nine faults for four different sensor suites. Here, basic four refers to ΔEGT, ΔWF, $\Delta N1$, and $\Delta N2$ sensors only, which are present in almost all operational gas turbine engines. The other results show the effect of the addition of $P25$ and $T25$ sensors between the LPC and the HPC, and $P3$ and $T3$ sensors before the burner.

With the four basic measurements, the average success rate is about 89%. However, there is considerable variation is the fault isolation accuracy for the different faults. In particular, the bleed faults and the LPC are not isolated well. In these cases, the bleed faults are sometimes confused with the LPC and vice versa. The confusion between the LPC module fault and the bleed fault is because of the similarity in the directions of the fingerprints, which can be seen in the fuzzy rules. In the cases where the random error is high, the fingerprints for the LPC look very similar to those of the bleeds and vice versa.

Placing additional sensors to measure $P25$ and $T25$ results in the fault isolation accuracy increasing from 86 to 95%. There is a marked increase in the isolation accuracy for the 2.5 and 2.9 bleed faults. In case additional sensors besides the basic four were used to measure $P3$ and $T3$, the average fault isolation accuracy increases from 89 to 96%. In particular, the isolation accuracy increased from 82 to 94% for the LPC fault. In addition, there is also an increase in the isolation accuracy for the 2.5 and 2.9 bleed faults. It can be seen that the $T3$ and $P3$ sensors result in a greater improvement in fault

TABLE 10.7

Comparison of Fault Isolation Accuracy (%), Neural Network, Fuzzy System, and Kalman Filter

Measurement Faults	Basic Four			All Eight		
	Fuzzy	Neural	Kalman	Fuzzy	Neural	Kalman
HPC	92	90	100	100	100	100
HPT	100	100	100	100	100	100
LPC	82	90	90	94	60	90
LPT	100	100	100	100	100	100
FAN	100	100	100	100	100	100
2.5 bleed	69	75	50	89	80	85
2.9 bleed	61	87	80	99	87	97
HPCSVM	100	100	100	100	100	100
P49ER	100	100	100	100	100	100
Average success rate	89	93	91	98	91	97

Source: Ganguli, R., *Journal of Propulsion and Power* 18(2):440–447, 2002. With permission.

isolation accuracy than the $T25$ and $P25$ sensors. However, when all eight sensors (basic four + $P25$ + $T25$ + $T3$ + $P3$) are used, the average fault isolation accuracy rises to 98%.

Table 10.7 compares results obtained using the fuzzy logic system with those obtained using neural network and Kalman filters in the previous two chapters. The same test cases were used for these results. There is a close agreement between the results. The fuzzy system and the Kalman filter give better results with the eight-sensor suite than with the neural network. This may be because the fuzzy system and the Kalman filter include the knowledge of the fingerprint charts in their rule base and influence coefficient matrix, respectively, whereas the neural network has to learn the relationships from the simulated training data.

The results discussed until now were obtained for test data generated using standard deviations given in Table 10.4 and that were also used in creating the membership functions for the FLS. Figures 10.6–10.10 show results obtained for the FLS for test data generated for various levels of uncertainty. Here σ_0 is the baseline standard deviation given in Table 10.4. Results are obtained for uncertainty ranging from 25% of the baseline value to 150% of the baseline value. Figure 10.6 shows the influence of measurement uncertainty for a system with only the four basic parameters. For very low uncertainty levels ($\sigma/\sigma_0 = 0.25$) the FLS shows 100% accuracy in fault isolation. As the measurement data deteriorate, the fault isolation success rate falls for the HPC, LPC, 2.5 bleed, and 2.9 bleed faults. The turbine and fan modules and HPCSVM and P49ER faults are isolated with 100% accuracy even with low-quality data.

The inclusion of $P25$ and $T25$ sensors results in some improvement in the fault isolation accuracy, as shown in Figure 10.7. Similarly, the inclusion of $P3$

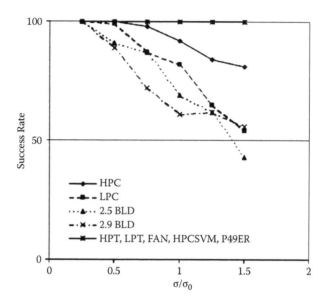

FIGURE 10.6
Fault isolation success rate with increasing uncertainty in data (basic four measurements only). (From Ganguli, R., *Journal of Propulsion and Power* 18(2):440–447, 2002. With permission.)

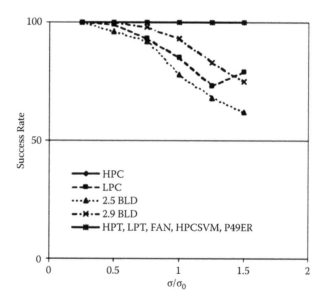

FIGURE 10.7
Fault isolation success rate with increasing uncertainty in data (basic four measurements + *P*25 + *T*25). (From Ganguli, R., *Journal of Propulsion and Power* 18(2):440–447, 2002. With permission.)

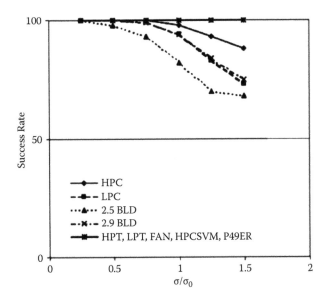

FIGURE 10.8
Fault isolation success rate with increasing uncertainty in data (basic four measurements + $T3$ + $P3$). (From Ganguli, R., *Journal of Propulsion and Power* 18(2):440–447, 2002. With permission.)

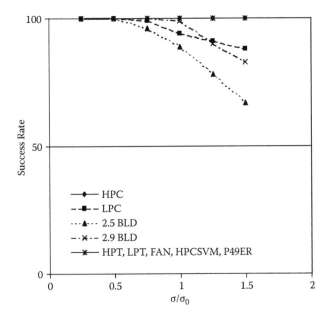

FIGURE 10.9
Fault isolation success rate with increasing uncertainty in data (all eight measurements). (From Ganguli, R., *Journal of Propulsion and Power* 18(2):440–447, 2002. With permission.)

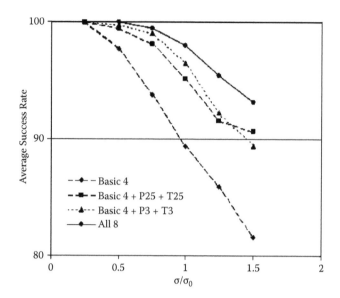

FIGURE 10.10
Comparison of average fault isolation success rate for different sensor suites and uncertainty in data. (From Ganguli, R., *Journal of Propulsion and Power* 18(2):440–447, 2002. With permission.)

and *T*3 sensors over and above the basic four results in some improvement in accuracy, as shown in Figure 10.8.

Finally, inclusion of all eight sensors shows considerable improvement in fault isolation accuracy, as shown in Figure 10.9. For uncertainty levels lower than σ_0, the fault isolation accuracy is significantly improved.

Thus, it is very important to focus on data cleaning and rectification methods, as well as improved sensors to remove potential outliers in the data. Recall that some of the first few chapters focused on such data clearing methods. The reader can now appreciate the importance of these methods. In addition, the robust nature of the FLS is clear from the fault isolation accuracy deteriorating gradually as uncertainty levels increase.

Finally, the average success rates for the four different measurement suites are summarized in Figure 10.10. Note that the importance of having the additional sensors becomes even more important as data quality falls. The FLS is able to identify the correct fault despite the presence of considerable uncertainty in the measurement.

The current study demonstrated the effectiveness of the fuzzy logic approach in the isolation of single faults following a sharp trend change. However, the study makes several assumptions and simplifications:

1. Unmodeled single faults, sensor faults, and multiple faults are not addressed.

2. Only module faults are considered and pure combustor performance problems are neglected.

3. The relationship between the module efficiencies and flow capacities/areas are considered fixed at the values provided by engine manufacturers.

4. The robustness of the system to noise in the measured data is analyzed by scaling all of the uncertainties to the same factor. The effect of scaling the uncertainty individually per measurement is neglected.

5. Missing measurements are not considered.

6. A "winner takes all" approach is used to find the most likely fault, and the possibility of using "beliefs" from the output of the system to study the possible confounding between the first and second most likely faults is not considered.

10.8 Summary

An FLS is developed for gas turbine engine performance diagnostics. It takes measurement deviations from a baseline model of four basic (cockpit) measurements (*EGT, WF, N1,* and *N2*) and analyzes deterioration in five modules (FAN, LPC, HPC, HPT, and LPT) and four system faults (2.5 bleed, 2.9 bleed, HPCSVM, P49Error). These measurements are available on most gas turbines. The FLS is based on fingerprint charts provided by engine manufacturers and widely used by airline engineers.

Results show that the FLS has a success rate of about 90% in isolating the faulty engine module with four cockpit measurements. In cases where the FLS is confounded, it was due to large uncertainty in the data. The FLS therefore can be used as a robust expert system for automating the process of interpreting gas turbine performance fingerprint charts.

Additional pressure and temperature sensors between the compressors (*P25* and *T25*) or before the burner (*P3* and *T3*) improve the fault isolation accuracy to about 95%. When all eight sensors are used together, the FLS shows an accuracy of 98% in fault isolation. Additional sensors become more important as data quality deteriorates. Therefore, additional sensors besides the four cockpit sensors are useful and recommended for accurate and robust fault isolation. It is shown that the fuzzy logic approach to gas turbine diagnostics is competitive with the Kalman filter approach.

11

Soft Computing Approach

In the last three chapters we focused on Kalman filters, neural networks, and fuzzy logic systems for fault isolation in gas turbines, respectively. Fuzzy systems are also universal function approximations in a manner similar to that in neural networks. Fuzzy systems also address the issue of uncertainty using a built-in fuzzifier, whereas a neural network learns the noise characteristics of the data through training. It was shown in Chapter 10 that fuzzy systems provide accurate fault isolation results for gas turbine diagnostics. However, the neural and fuzzy methods for diagnostics are highly configuration dependent, meaning that if the underlying model used to obtain fault signatures or the measurement uncertainties of the signal changed, the diagnostic systems have to be redeveloped. Since there are many different engines operating with different airlines, there are likely to be many possible combinations of fault signatures and measurement uncertainties for the fault isolation systems that need to be developed. Very often the process of redeveloping the underlying numerics or rules for the diagnostic system is a trial and error process that can be very tedious and requires considerable human effort.

In this chapter, we describe a genetic fuzzy system [106–108] that allows for easy development of the rule base for an engine given fault signature and measurement uncertainties [109]. Unlike conventional fuzzy logic applications, where rules are generated based on operators' experience or general knowledge of the system in a heuristic way, in such a system, optimization techniques such as genetic algorithms are used to tune the fuzzy membership functions and rules. Typically, if Gaussian fuzzy sets are used, the number of fuzzy sets, their midpoints, and standard deviations can be used as design variables. Genetic algorithms are used to maximize the performance of a fuzzy system through automatically selecting the number of fuzzy sets and membership functions based on the fault signatures of the engine and measurement uncertainties to achieve the goal of minimizing the number of design variables. The genetic fuzzy system thus automates the creation of the fuzzy system, greatly reducing the human effort needed. Furthermore, a radial basis function neural network (RBFNN) preprocessor is used for denoising signals typical of path measurements. The advantages of using such a signal processing algorithm prior to fault isolation by a genetic fuzzy system are shown.

11.1 Gas Turbine Fault Isolation

A twin-spool gas turbine is shown in Figure 1.2 with five modules, among which are the FAN, LPC, and HPC modules, which have efficiencies and the flow capacities associated with them, and the HPT and LPT modules, which have efficiencies and areas associated with them. The fingerprints or fault signatures relating a change in measurement deltas for four basic parameters with the faulty module are given in Table 3.1. The four basic parameters found in almost all engines are exhaust gas temperature (EGT), low rotor speed ($N1$), high rotor speed ($N2$), and fuel flow (WF). The fault signatures in Table 3.1 assume the following couplings between module efficiencies and flow capacities:

1. FAN   Coupled FAN (-2% η, -2.5 FC)
2. LPC   Coupled LPC (-2% η, -2.2% FC)
3. HPC   Coupled HPC (-2% η, -1.6 FC)
4. HPT   Coupled HPT (-2% η, -1.5 $FP4$)
5. LPT   Coupled LPT (-2% η, $+3.3\%$ $FP45$)

Here FC is the flow capacity, $FP4$ is the high-pressure turbine area, and $FP45$ is the low-pressure turbine area. Each fault is modeled as a 2% decrease in efficiency from the baseline good engine. Since the fault signatures are derived from influence coefficients, they are only approximately correct because they do not account for uncertainties in the measurement process. Each gas path measurement is associated with an uncertainty. One measure of this uncertainty is the standard deviations from revenue service data. As given in the previous chapters, typical standard deviations for ΔEGT, $\Delta N1$, $\Delta N2$, and ΔWF are 4.23°C, 0.25%, 0.17%, and 0.50%, respectively.

11.2 Neural Signal Processing—Radial Basis Function Neural Networks

Since gas turbine measurements are often contaminated with noise and outliers, it is useful to perform a data cleaning function prior to fault isolation. In this chapter, we use a radial basis function neural network (RBFNN) for removing noise from simulated signals. Radial basis networks are an alternative to the more widely used multilayer perceptron networks trained using the back-propagation algorithm and take much less computer time for training [110–112].

The RBFNN model consists of three layers: an input layer, a hidden (kernel) layer, and an output layer. The nodes within each layer are fully connected to the previous layer. The input variables are each assigned to a node in an input layer and pass directly to the hidden layer without weights. The hidden nodes or units contain the RBFs, also called transfer functions. An RBF is symmetrical about a given mean or center point in a multidimensional space. In the RBFN, a number of hidden nodes with RBF activation functions are connected in a feed-forward parallel architecture. The parameters associated with the RBFs are optimized during training. These parameter values are not necessarily the same throughout the network, nor are they directly related to or constrained by the actual training vectors. When the training vectors are presumed to be accurate, i.e., nonstochastic, and it is desirable to perform a smooth interpolation between them, then linear combinations of RBFs can be found that give no error at the training vectors. The methods of fitting RBFs to data, for function approximation, are closely related to distance weighted regression. The RBF expansion for one hidden layer and an arbitrary RBF is represented by the equation

$$y_k(x) = \sum_{i=1}^{H} w_{ki} \exp(-\|c_i - x\|/\sigma_i^2) \tag{11.1}$$

where y_k = kth output, w_{ki} = weight from the ith kernel node to the kth output node, c_i = centroid of the ith kernel node, σ_i = width of the ith kernel node, and H = number of kernel nodes. The parameters of the RBF w_{ki}, c_i, and σ_i are commonly chosen by first selecting randomly or uniformly the c_i and then using singular value decomposition (SVD) to solve for w_{ki} and σ_i. This approach is not the most satisfactory. A better approach involves using K-means clustering to determine the c_i, a K-nearest heuristic to determine the σ_i, and multiple linear regressions to determine the w_{ki}. The K-means clustering algorithm finds a set of cluster centers and a partition of the training data into subsets. Each cluster center is then associated with one of the H kernels or centers in the hidden layer. After the centers are established, the width of each kernel is determined to cover the training points to allow a smooth fit of the desired network outputs.

11.3 Fuzzy Logic System

A fuzzy logic system (FLS) is a nonlinear mapping of an input feature vector into a scalar output [101]. A typical FLS maps crisp inputs to crisp outputs using four basic components: rules, fuzzifier, inference engine, and defuzzifier. Once the rules driving the FLS have been fixed, the FLS can be expressed as a mapping of inputs to outputs. The output of the FLS is kept as fuzzy sets, as they are easier to interpret linguistically for diagnostic and prognostic action. Rules for the fuzzy system are obtained by fuzzification of the numerical values in the fingerprint charts using the following procedure:

Algorithm 11.1

1. Each measurement delta is divided into N fuzzy sets whose geometry is selected by the designer.
2. A set of four measurement deltas corresponding to a given module fault is input to the FLS, and the degree of membership of the elements of ΔEGT, ΔWF, $\Delta N2$, and $\Delta N1$ is obtained.
3. Each measurement delta is then assigned to the fuzzy set with the maximum degree of membership.
4. One rule is obtained for each module fault by relating the measurement deltas with a maximum degree of membership to a module fault. ■

For any given input set of measurement deltas, the fuzzy rules are applied using product implication. Once the fuzzy rules are applied for a given measurement, we have degree of membership for FAN, LPC, HPC, HPT, and LPT. For fault isolation, we are interested in the most likely fault. The fault with the highest degree of membership is selected as the most likely fault.

The main problem in Algorithm 11.1 is in the selection of the number and type of fuzzy sets in step 1. Typically, designers select the number and geometry of the fuzzy sets based on knowledge of the problem. For example, the measurements may be classified into five fuzzy sets named very low, low, medium, high, and very high. In case Gaussian functions are selected as membership functions, the midpoints and standard deviations associated with each Gaussian fuzzy set need to be selected so that the entire measurement range is spanned by the fuzzy sets and there is some intersection between the sets. Thus, the designer must manually iterate over Algorithm 11.1 to obtain a fuzzy system that has good performance.

11.4 Genetic Algorithm

Genetic algorithms (GAs) are a probabilistic search method [113–115]. A brief introduction to GA is given below. The genetic algorithm is motivated by the hypothesized natural process of evolution in biological populations, where genetic information stored in chromosomal strings evolves over generations to adapt favorably to a static or changing environment. The algorithm is based on elitist reproduction strategy, where members of the population, who are deemed most fit, are selected for reproduction and are given the opportunity to strengthen the chromosomal makeup of progeny generation. This approach is facilitated by defining a fitness function or a measure indicating the goodness of a member of the population in the given generation during the evaluation process.

To represent designs as chromosome-like strings, the design variable is converted to its binary equivalent and thereby mapped into a fixed-length string of 0s and 1s. A number of such strings constitute a population of designs, with each design having a corresponding fitness value. This fitness value could be the objective function $F(X)$ for a function maximization problem. Thus, the GA can be used to solve optimization problems of the form:

Maximize $F(X)$

Subject to $X_i^{(\min)} \leq X_i \leq X_i^{(\max)}$

The starting population is selected randomly in the domain lying between the minimum and maximum values of X, and then the following genetic operators are applied to improve results.

1. *Reproduction.* Individuals are selected and the probability of selection is based on their fitness value. The new population pool has higher average fitness value than the previous pool.

2. *Crossover.* In the two-point crossover approach, two mating parents are selected at random; the random number generator is invoked to identify two sites on the strings, and the strings of 0s and 1s enclosed between the chosen sites are swapped between the mating strings.

3. *Mutation.* A few members from the population pool are taken according to the probability of mutation p_m, and a 0 to 1 or vice versa is switched at a randomly selected mutation site on the chosen string.

The process of reproduction, crossover, and mutation constitutes one generation of the GA. After several generations, the GA is stopped and the best point among the values taken as the optimal point. Being a probabilistic search method, GAs are very good at finding global maxima. Furthermore, GAs need only function values and not gradient information, which makes them easy to use for real systems where accurate gradient information is difficult to obtain, and local minima may occur. However, they are computationally expensive.

11.5 Genetic Fuzzy System

There are two main problems in the generation of fuzzy systems [106]. The first is that it is difficult to select the appropriate number of fuzzy sets. The second is selection of the membership functions. The rules need to be created for a given number of fuzzy sets and type of membership functions. However, if the number of fuzzy sets or type of membership function changes, the rules can change. Therefore, any change in the membership functions or the number of fuzzy sets leads to a change in the rule base; the process of designing a fuzzy system is iterative and can become very cumbersome for a human designer. It is therefore desirable to create an automated procedure for the design of fuzzy systems. A genetic algorithm is used to facilitate the design of the fuzzy system. The approach is discussed below:

Algorithm 11.2

1. Define maximum and minimum values for a measurement delta Δz by $\Delta z^{(max)}$ and $\Delta z^{(min)}$, respectively.
2. Define the universe of discourse for Δz to be the set of real numbers between the minimum and maximum values, $U(\Delta z) = [\Delta z^{(min)}, \Delta z^{(max)}]$.
3. Define $L(\Delta z) = \Delta z^{(max)} - \Delta z^{(min)}$ as the length of the universe of discourse.
4. Divide U into N Gaussian fuzzy sets $F_1, F_2, ..., F_N$ and define the midpoint of fuzzy point F_1 by $\Delta z^{(min)}$ and of fuzzy set F_N by $\Delta z^{(max)}$, respectively. These fuzzy sets can be defined using the following equation:

$$\mu(x) = e^{-0.5\left(\frac{x-m}{\sigma}\right)^2} \tag{11.2}$$

where m is the midpoint of the fuzzy set and σ is the uncertainty (standard deviation) associated with the variable.

5. Assuming the fuzzy sets are equally spaced, calculate the midpoints of fuzzy set F_2 as $\Delta z^{(min)} + \Delta m$, set F_3 as $\Delta z^{(min)} + 2*\Delta m$, and set F_i as $\Delta z^{(min)} + (i-1)\Delta m$, where

$$\Delta m = \frac{L(\Delta z)}{N-1} \tag{11.3}$$

6. Allow the fuzzy sets for the measurement delta Δz to move together along the number line by an amount x. This allows the midpoints of the fuzzy sets to change, along with the values $\Delta z^{(min)}$ and $\Delta z^{(max)}$.

However, the distance $L(\Delta z)$ remains constant. With this definition, the midpoints of the fuzzy sets are defined once N and x are selected. Select the standard deviation of the fuzzy set for measurement Δz as the measurement uncertainty of Δz. ∎

Algorithm 11.3

1. Define the maximum and minimum values for each measurement ΔEGT, $\Delta N1$, $\Delta N2$, and ΔWF from the fault signatures as shown in Table 2.1. Thus for ΔEGT, the maximum and minimum values are 21.77 and $-7.72°C$, respectively.
2. Define the range spanned by each variable as $L_1 = L(\Delta EGT)$, $L_2 = L(\Delta N1)$, $L_3 = L(\Delta N2)$, and $L_4 = L(\Delta WF)$.
3. Choose N fuzzy sets to partition each measurement. To start the algorithm, use $N = 2$.
4. Let x_1, x_2, x_3, and x_4 define the tuning variables associated with ΔEGT, $\Delta N1$, $\Delta N2$, and ΔWF, respectively. To start the algorithm, select random values satisfying $-25\%\ L_i \leq x_i \leq 25\%\ L_i$, $i = 1, 4$. Choose σ for ΔEGT, $\Delta N1$, $\Delta N2$, and ΔWF as $4.23°C$, 0.25%, 0.17%, and 0.50%, respectively.
5. Generate the fuzzy system from the numerical data using the conventional procedure outlined in Algorithm 11.1.
6. Using a sample of 100 noisy data points, calculate the success rate as

$$S = 100\frac{N_C}{N_T}$$

where N_C is the number of correct classifications and N_T is the total number of classifications.
7. Use GA to solve the optimization problem by taking the best solution from $N_{gen}^{(max)}$ generations:
   Maximize $S(x_1, x_2, x_3, x_4)$.
   Subject to $-25\%\ L_i \leq x_i \leq 25\%\ L_i$, $i = 1, 4$.
8. Increase N by 1:
   a. If $N < N^{(max)}$, g 0 to 3.
   b. Else select N with highest success rate S (if highest S is obtained by more than one value of N, select the lowest N that gives the highest S). ∎

The only values that need to be the input to the genetic fuzzy system (GFS) are the values of measurement deltas corresponding to each fault, and the fault signature based on the linearized influence coefficients at the current operating point. For the standard deviations of the Gaussian fuzzy sets, we use the measurement uncertainty data that can be obtained by a statistical analysis of engine data. If the measurement uncertainties change, the GFS can be tuned to the different numerics. Thus, we get an automatic system that greatly reduces the need of manual manipulation.

11.6 Numerical Simulations

In this chapter, a maximum of nine generations of the GA are used for each N value of the fuzzy sets. The population size, crossover probability, and mutation probability are chosen as 20, 0.8, and 0.1, respectively. The maximum number of fuzzy sets is selected as 10.

Since genetic algorithms are computationally intensive, the issue about computation time is important for practical implementation. As an example, the code implementing the algorithm in this chapter takes about 3–5 minutes to run on MATLAB® on a PC with the full nine generations of GA. However, in many cases, the convergences occur in two or three generations given that we use only four design variable has a starting population of 20. Each design variable is represented by a 10-bit string.

As mentioned earlier, a standard approach in the design of the optimal fuzzy system is to consider the midpoints and standard deviations of each fuzzy set as design variables. If there are N fuzzy sets and M measurements, the maximum number of midpoint design variables is $N*M$, and the maximum number of standard deviation design variables is $N*M$. The total number of design variables is therefore $2*N*M$. For the case with $N = 6$ and $M = 4$, we would have a total of $2*6*4 = 48$ design variables, leading to high computer time requirements.

The algorithm in this chapter uses some prior knowledge of the problem to reduce the number of design variables dramatically. The standard deviations are thus selected to be equal to the measurement uncertainties. In this manner, the fuzzifier is able to act as a filter that addresses noise in the data in a direct manner. By making the requirement that the universe of discourse only spans the neighborhood of the measurements, the region where fuzzy set discretization is needed is optimized. Using a uniform distribution of fuzzy sets leads to so-called design variable linking in optimization and allows the midpoints to be defined using only two variables for each measurement: the number of fuzzy sets N and the translation variable x. For a given number of fuzzy sets, the number of design variables is equal to the number of measurements, which is four in this case. The fuzzy system is tested using simulated data developed from the fault signatures shown in Table 3.1. For each module, 100 noisy data sets are generated for module faults with 2% deterioration in efficiency. Noise is added to the simulated measurement deltas using the typical standard deviations for ΔEGT, $\Delta N1$, $\Delta N2$, and ΔWF as 4.23°C, 0.25%, 0.17%, and 0.50%, respectively.

Figure 11.1 shows the success rate for the optimal GFS as the number of fuzzy sets is increased from 2 to 9. For each value of N in this figure, the optimal values of x are calculated using Algorithm 11.3. For only two fuzzy sets, the success rate is about 80% and quickly rises as the number of sets increases. The number $N = 6$ is selected by Algorithm 11.3 as the point where the GFS is optimal with a minimum number of sets. Figure 11.2 shows

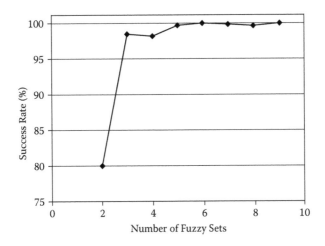

FIGURE 11.1
Change in fault isolation success rate for increasing number of fuzzy sets. (From Verma, R., et al., *Applied Mathematics and Computation* 172(2):1342–1363, 2006. With permission.)

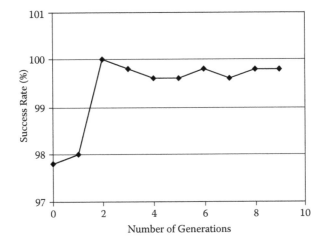

FIGURE 11.2
Evolution of success rate for a fuzzy system with six sets with generation of genetic algorithm. (From Verma, R., et al., *Applied Mathematics and Computation* 172(2):1342–1363, 2006. With permission.)

the success rate of the fuzzy system with six sets as the GA generations progress. In this case, only two generations were needed to achieve a success rate of 100%, and the values of x corresponding to the second generation of GA are selected by Algorithm 11.3 as the optimal fuzzy system.

Tables 11.1–11.5 provide the midpoints of the fuzzy sets for the four measurements as the number of fuzzy sets increases from two to six. The starting values in Table 11.1 show two fuzzy sets with midpoints centered near the maximum and minimum values of the measurements. The values in Table 11.5 correspond to the case where $N = 6$ in Figure 11.1 and $N_{gen} = 2$ in Figure 11.2.

Figure 11.3 shows the starting case with two fuzzy sets, which is a crude discretization. Figures 11.4, 11.5, and 11.6 show the cases with three, four, and five fuzzy sets, respectively. In Figure 11.7, the optimal level of discretization with six fuzzy sets is achieved.

TABLE 11.1

Midpoints of Two Fuzzy Sets

ΔEGT (°C)	−7.69	21.8
$\Delta N1$ (%)	−1.93	1.38
$\Delta N2$ (%)	−1.1	1.3
ΔWF (%)	−1.89	2.61

Source: Verma, R., et al., *Applied Mathematics and Computation* 172(2):1342–1363, 2006. With permission.

TABLE 11.2

Midpoints of Three Fuzzy Sets

ΔEGT (°C)	−8.16	6.58	21.33
$\Delta N1$ (%)	−2.4	−0.75	0.91
$\Delta N2$ (%)	−1.57	−0.37	0.83
ΔWF (%)	−2.36	−0.11	2.14

Source: Verma, R., et al., *Applied Mathematics and Computation* 172(2):1342–1363, 2006. With permission.

TABLE 11.3

Midpoints of Four Fuzzy Sets

ΔEGT (°C)	−8.31	1.52	21.18	11.35
$\Delta N1$ (%)	−2.55	−1.45	0.76	−0.35
$\Delta N2$ (%)	−1.72	−0.92	0.68	−0.12
ΔWF (%)	−2.51	−1.01	1.99	0.49

Source: Verma, R., et al., *Applied Mathematics and Computation* 172(2):1342–1363, 2006. With permission.

TABLE 11.4

Midpoints of Five Fuzzy Sets

ΔEGT (°C)	−7.82	−0.44	6.92	14.3	21.67
ΔN1 (%)	−2.06	−1.23	−0.4	0.42	1.25
ΔN2 (%)	−1.23	−0.63	−0.03	0.57	1.17
ΔWF (%)	−2.02	−0.89	0.23	1.36	2.48

Source: Verma, R., et al., *Applied Mathematics and Computation* 172(2):1342–1363, 2006. With permission.

TABLE 11.5

Midpoints of Six Fuzzy Sets

	VL	L	ML	MH	H	VH
ΔEGT (°C)	−9.62	−3.72	2.17	8.07	13.97	19.87
ΔN1 (%)	−2.23	−1.56	−0.9	−0.24	0.42	1.08
ΔN2 (%)	−1.21	−0.72	−0.25	0.23	0.71	1.19
ΔWF (%)	−2.25	−1.35	−0.45	0.45	1.35	2.25

Source: Verma, R., et al., *Applied Mathematics and Computation* 172(2):1342–1363, 2006. With permission.

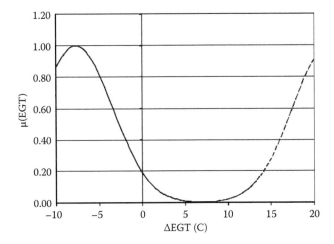

FIGURE 11.3

Discretization of universe of exhaust gas temperature using two fuzzy sets. (From Verma, R., et al., *Applied Mathematics and Computation* 172(2):1342–1363, 2006. With permission.)

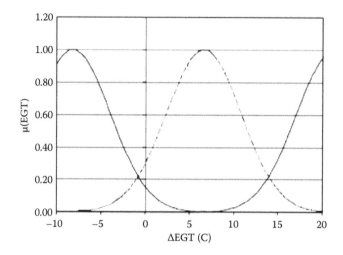

FIGURE 11.4
Discretization of universe of exhaust gas temperature using three fuzzy sets. (From Verma, R., et al., *Applied Mathematics and Computation* 172(2):1342–1363, 2006. With permission.)

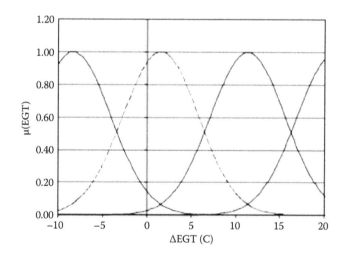

FIGURE 11.5
Discretization of universe of exhaust gas temperature using four fuzzy sets. (From Verma, R., et al., *Applied Mathematics and Computation* 172(2):1342–1363, 2006. With permission.)

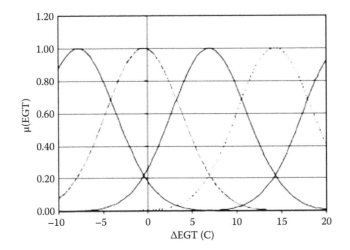

FIGURE 11.6
Discretization of universe of exhaust gas temperature using five fuzzy sets. (From Verma, R., et al., *Applied Mathematics and Computation* 172(2):1342–1363, 2006. With permission.)

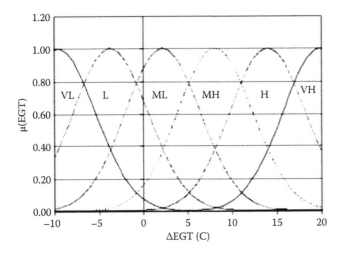

FIGURE 11.7
Discretization of universe of exhaust gas temperature using six fuzzy sets.

In Table 11.5 each fuzzy set is assigned a linguistic value of very low (VL), low (L), medium-low (ML), medium-high (MH), high (H), and very high (VH). These linguistic measures are shown in Figure 11.7 for the six ΔEGT fuzzy sets. The fuzzy rule base for the case with six fuzzy sets is shown in Table 11.6. Table 11.6 is the result of fuzzification of the numerical data in Table 2.1. These rules can be read as follows for the FAN module:

IF

> ΔEGT is very low AND
>
> $\Delta N1$ is very high AND
>
> $\Delta N2$ is low AND
>
> ΔWF is very low

THEN

> Problem in FAN module

The rules for the other modules can be similarly interpreted. These rules provide a knowledge base and represent how a human engineer would interpret data to isolate an engine fault using fingerprint charts.

Table 11.7 shows the success rate of the fuzzy set with 100 noisy data points. The noisy data points for testing are different from data used for developing the rule base of the fuzzy system. The average success rate is 100%, compared to 98.2% for the manually designed fuzzy system discussed in Chapter 10. The manually designed fuzzy system showed some problems in differentiating between faults in the LPC and those in the HPC. It is clear that GFS is able to identify the correct fault despite the presence of considerable uncertainty in measurements.

The effect of noise on the GFS is shown in Figure 11.8, and the results are compared with data from the fuzzy system from Chapter 10. Here the noise ratio is defined as σ/σ_0, where σ_0 is the baseline noise level used for developing the GFS and σ is the noise level in the simulated data used for testing. It is clear that both systems show a decline in the average fault isolation success rate with increasing noise levels in the data. However, the GFS appears to

TABLE 11.6

Rules for Optimal Fuzzy System with Six Fuzzy Sets

	ΔEGT	$\Delta N1$	$\Delta N2$	ΔWF
FAN	VL	VH	L	VL
LPC	ML	MH	H	MH
HPC	MH	MH	ML	H
HPT	VH	MH	VL	VH
LPT	L	VL	VH	VL

Source: Verma, R., et al., *Applied Mathematics and Computation* 172(2):1342–1363, 2006. With permission.

TABLE 11.7

Results for Optimal Fuzzy System and Manually Designed System

Module	Success Rate (%) (Optimal)	Success Rate (%) (Manually Designed)
HPC	100	94
HPT	100	100
LPC	100	97
FAN	100	100
LPT	100	100
Average success rate	100	98.2

Source: Verma, R., et al., *Applied Mathematics and Computation* 172(2):1342–1363, 2006. With permission.

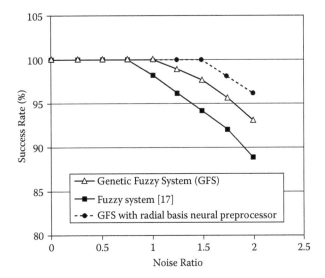

FIGURE 11.8
Success rate in fault isolation with increasing noise levels in data.

show a somewhat better performance as the noise level increases. This is due to the optimal nature of the fuzzy system developed and the use of formal optimization methods rather than a trial and error process in maximizing the success rate. The result of applying a neural network preprocessor to the GFS is discussed below.

To study the signal processor, we assume time series of 100 discrete points. From $k = 0$ to $k = 50$, the signal changes linearly from 0 to sign (Δz) $\sigma_0/2$. From $k = 50$ to $k = 51$, the signal changes by Δz. From $k = 51$ to 100 the signal changes from Δz to Δz + sign (Δz) $\sigma_0/2$. This simulates a single-fault situation, where a step jump equal to the measurement deltas corresponding to the module faults is added to a linearly varying signal. As an example, the ΔEGT

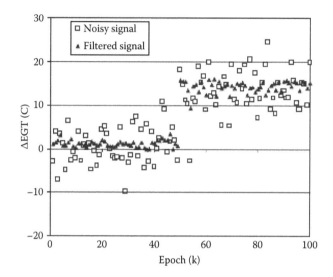

FIGURE 11.9
Noisy and filtered ΔEGT signal simulating HPC fault.

variation for an HPC fault is simulated using a linear variation from 0°C at $k = 1$ to $4.23/2 = 2.115$°C at $k = 50$, followed by a change to $13.6 + 4.23/2 = 15.715$°C at $k = 51$, and a linear variation thereafter to $13.6 + 4.23 = 17.83$°C. Figure 11.9 shows the noisy signal and RBF filtered signal.

For determining the RBF unit centers, we use a K-means clustering algorithm. The K-means clustering algorithm finds a set of clusters each with centers from the given training data. The cluster centers become the centers of the RBF units. The number of clusters is a design parameter and determines the number of RBF units, i.e., nodes, in the hidden layer. We have used $H = 20$. When the RBF centers have been established, the widths of each RBF can be calculated.

The width of any RBF distance to the nearest p RBF units, where p is a design parameter for the RBFN, for unit t is given by

$$\sigma_i = \sqrt{\left[\frac{1}{p} \sum_{j=1}^{p} \sum_{k=1}^{r} (x_{\hat{k}i} - x_{\hat{k}j})^2 \right]} \qquad (11.4)$$

where $x_{\hat{k}i}$ and $x_{\hat{k}j}$ are the kth entries of the centers of the ith and jth hidden units. We have used $p = 5$. When the centers and widths of the RBF units have been chosen, then the $N = 100$ training samples are processed through the hidden nodes to generate an $H \times N$ matrix, called A. Let T be the $M \times N$ desired output matrix for the training patterns and $M = 100$ the number of output nodes. The objective is to find the weights that minimize the error between the actual output and the desired output of the network. Essentially, we are trying to minimize the objective (cost) function

$$\|T - WA\|$$

where W is the $M \times H$ matrix of weights on the connections between the hidden and output nodes of the network. We train the RBF network with added Gaussian noise at $\sigma_0 = 4.23°C$, 0.25%, 0.17%, and 0.50%, respectively, for ΔEGT, $\Delta N1$, $\Delta N2$, and ΔWF.

Noise is added to the ideal signal using a baseline value σ_0 of typical standard deviations for ΔEGT, $\Delta N1$, $\Delta N2$, and ΔWF as 4.23°C, 0.25%, 0.17%, and 0.50%, respectively. The filtered signals in Figures 11.9 and 11.10 show considerable noise reduction while preserving the nature of the step edge. These data represent one noisy signal for each measurement. The visual quality of the data is considerably improved. Similar results are obtained for all the signals corresponding to the faults in Table 2.1. To summarize these results concisely, the following noise reduction measure is defined based on the mean absolute error (*MAE*) criteria:

$$MAE^{(noisy)} = \sum_{i=1}^{N} \frac{1}{N} \left| \Delta z_i^{(noisy)} - \Delta z_i^{(ideal)} \right| \qquad (11.5)$$

$$MAE^{(filtered)} = \sum_{i=1}^{N} \frac{1}{N} \left| \Delta z_i^{(filtered)} - \Delta z_i^{(ideal)} \right| \qquad (11.6)$$

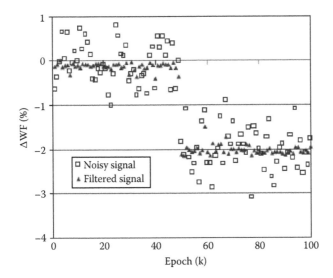

FIGURE 11.10
Noisy and filtered ΔWF signal simulating HPC fault.

TABLE 11.8

Noise Reduction Using Radial Basis Neural Network

	ΔEGT (°C)	$\Delta N1$ (%)	$\Delta N2$ (%)	ΔWF (%)
HPC	78.84	67.03	67.38	81.87
HPT	84.24	72.07	83.38	83.71
LPC	77.5	74.34	78.48	77.95
FAN	74.8	82.43	79.04	80.62
LPT	68.76	83.83	84.68	82.95
Average	76.83	75.94	78.59	81.42

Source: Verma, R., et al., *Applied Mathematics and Computation* 172(2):1342–1363, 2006. With permission.

$$N_R = 100 \frac{MAE^{(noisy)} - MAE^{(filtered)}}{MAE^{(noisy)}} \tag{11.7}$$

For each signal, 100 samples of noisy test data are created and the noise reduction calculated. These values are summarized in Table 11.8 and show a noise reduction averaging between 75 and 81%. Results in this chapter clearly demonstrate the power of the soft computing framework for automated decision making under uncertainty. The approach uses the concept of hybridization in soft computing, where using different techniques such as neural networks, genetic algorithms, and fuzzy logic together gives better results than if each method were used individually. The hybridization process uses the strengths of each different approach to attack the problem.

11.7 Summary

A novel genetic fuzzy system (GFS) is discussed in this chapter for fault isolation in gas turbine engines. The GFS has better performance than a manually designed fuzzy system because GFS automatically selects the number of fuzzy sets and membership functions based on the fault signatures of the engine and measurement uncertainties. This minimizes the computational demand of the model generation and allows problems with realistic dimensions to be considered. The fault signatures are derived from influence coefficients. A radial basis function neural network (RFBNN) is also studied for data cleaning prior to fault isolation. RBFNN shares the universal approximation capability, and takes much less training time and offers much better performance than the traditional linear filter.

For simulated faults considered in this chapter, the GFS achieved a success rate of 100% for the five module faults (HPC, LPC, FAN, HPT, and LPT) and four measurements (ΔEGT, $\Delta N1$, $\Delta N2$, and ΔWF). In contrast, a manually developed fuzzy system achieved a success rate of 98% with some confounding between the LPC and HPC module faults.

The trial and error process used to design a fuzzy system leads to considerable human labor and is often suboptimal. Different turbine engines can have different numerics, such as influence coefficients and measurement uncertainties, and it is a tedious process to develop a fuzzy system for each case.

The GFS automates the process of design of the fuzzy system. As noise levels in data increase, the GFS retains its edge over the manually designed fuzzy system, giving a 2–5% higher success rate with the same numerics.

A radial basis neural network prefilter achieved 75–81% noise reduction for simulated signals with linear deterioration and step changes. When the neural network is used to prefilter signals prior to fault isolation, the accuracy of the GFS is further improved for lower-quality data by 2–4%.

12

Vibration-Based Diagnostics

The previous chapters have considered diagnostics based on gas path measurements. In this chapter, we look at a problem where vibration characteristics are used for gas turbine diagnostics. The present chapter focuses on turbine blade damage. Turbine blades undergo cyclic loading causing structural deterioration, which can lead to failure. It is important to know how much damage has taken place at any particular time to monitor the condition or health of the blade and to avoid any catastrophic failure of the blades. Several studies look at damage at a given time during the operational history of the structure. This is typically called diagnostics and involves detection, location, and isolation of damage from a set of measured variables. The detection function is most fundamental, as it points out if the damage is present or not. However, some level of damage due to microcracks and other defects is always present in a structure. The important issue of indicating when to detect damage depends on how much life is left in the structure. It is not advantageous to detect small levels of damage in a structure. It would be useful if damage detection were triggered some time before final failure.

The subject of prognostics involves predicting the evolution of structural or vibration characteristics of the system with time and is important for prediction of failure due to operational deterioration. Some recent studies have considered dynamical systems approaches to model damage growth based on differential equations [116], while others have used physics-based models [117].

The stiffness of the structure is gradually reduced with crack growth, and stiffness is related to the vibrational characteristics of the structure. The decreased frequency shows that stiffness of the structure is decreasing, and thus serves as a damage indicator for monitoring crack growth in the structure. Selected studies have looked at modeling turbine blades as rotating Timoshenko beams with twist and taper [118–120]. Some studies have addressed damage in such beams using vibrational characteristics [121, 122]. However, these studies typically address damage at a given time point in the operational history and do not look at the effect of damage growth on the vibrational characteristics. In additions, turbine blades are designed to sustain a considerable amount of accumulated damage prior to failure. Therefore, it is desirable to indicate that a blade is damaged at the point when its operational life is almost over.

To study the structural dynamic behavior of the damaged beam, a damage model needs to be integrated into the finite element analysis. A number of damage models are available in [123]. Some use a fracture mechanics-based approach to include a crack in the finite element formulation [124]. The effect of damage growth can be modeled using crack growth models such as the Paris law [125]. Another approach is to use a phenomenological model devised from experimental measurements, which relates the loss of stiffness at the continuum level of the structure to the growth of damage. For example, most steam turbine rotor fractures are caused by low cycle fatigue (LCF) [126]. The high strains that cause LCF typically occur during the cold start and sliding parameter stop phases of steam turbine operation. Due to high-strain rates implicit in LCF, it is typically accompanied by plastic deformation. LCF is also the most significant life-reducing failure mechanism for military aircraft engines due to the throttle changes experienced during operation [127]. Therefore, the study of LCF effect on turbine blades is an important problem.

Most work on LCF is based on linear damage accumulation theory [126]. However, this theory results in crude damage accumulation calculations, and the predicted results are quite different in practice. In reality, LCF damage accumulation is a nonlinear process. Continuum damage mechanics (CDM) provides a way to describe the entire failure process from microrupture initiation and visual crack formation to structure failure. A damage model based on CDM for LCF damage analysis is chosen in this chapter. LCF occurs after a relatively low number of cycles (typically <10,000) of high-stress amplitude. The CDM models are easier to include in finite element analysis. They permit the finite element analysis and damage growth analysis to be effectively decoupled.

The CDM models use a damage variable $D = 1 - E/E_0$, which ranges from $D = 0$ for the undamaged material to $D = 1$ for complete failure. The CDM models the evolution of D from 0 to 1. Since it is very difficult to measure D in an operational turbine blade, another damage indicator that is dependent on D needs to be used. Frequency is one such indicator, and several methods have been developed in recent years for frequency measurement of turbine blades.

The simplest method of determining frequencies is by using blade-mounted strain gauges. Typically, blade mounted gauges can be used to detect the blade tip deflection. Lawson and Ivey [128] mention that two strain gauges per blade can be used to detect the first four modes of blade vibration. They developed an electronic circuit to energize the blade-mounted strain gauges and amplify the resulting signal. This system was developed for on-rotor operation.

Another promising approach for determining vibration characteristics of rotating engine blades is the use of blade tip timing methods. These methods have evolved because of the disadvantages of blade-mounted strain gauges, which include a complex installation procedure, a limited number of gauges, and the possibility of failure of the strain gauges due to high temperature experienced in the turbines. Strain gauges can also require complex telemetry of slip ring systems and can interface with the mechanical properties of the bladed assembly [129]. Tip timing methods typically use optical probes mounted on the engine casing and can measure the motion of each blade. These probes measure the time of arrival (TOA) of each blade relative to a once-per-revolution (OPR) or multiple-per-revolution (MPR) probe mounted on the rotor shaft. The difference between the TOA of a vibrating blade and the computed TOA of a nonvibrating blade provides the raw data that is used to determine the instantaneous blade displacement. Optical systems can be used to obtain vibration amplitude using curve fitting and frequencies using Fourier analysis. According to Carrington et al. [129], tip timing data acquisition is now in its fourth generation and is sufficiently advanced. Very recent work has also looked at the potential of a dual-use capacitance probe sensor to measure tip timing and tip clearance as alternatives to optical devices [128]. It is therefore possible to measure the frequencies of turbine blades quite accurately using the tip timing method.

The present chapter incorporates a CDM-based damage model for LCF damage into the finite element analysis of a rotating turbine blade. Numerical results are obtained to track the changes in frequencies with damage growth. A method of detecting the final stage of damage based on placing thresholds on the frequency change is proposed. The material discussed in this chapter was first proposed by Kumar, Roy, and Ganguli in their paper [130].

12.1 Formulations

12.1.1 Modeling of Turbine Blade

The turbine blade is modeled as a tapered, twisted, and rotating Timoshenko beam. The geometry of the beam is defined in Figure 12.1(a), the degrees of freedom for the element in Figure 12.1(b), the cross section in Figure 12.1(c), and the finite element used is shown schematically in Figure 12.1(d).

The total strain energy U of a beam of length l, due to bending and shear deformation (in-plane and out-of-plane directions), including rotary inertia and rotation effects, is given by

$$
\begin{aligned}
U = \int_0^l &\left[\left\{\frac{EI_{xx}}{2}\left(\frac{\partial^2 w_b}{\partial z^2}\right)^2 + EI_{xy}\frac{\partial^2 w_b}{\partial z^2}\frac{\partial^2 v_b}{\partial z^2} + \frac{EI_{yy}}{2}\left(\frac{\partial^2 v_b}{\partial z^2}\right)^2\right\}\right. \\
&\left. + \frac{\mu AG}{2}\left\{\left(\frac{\partial w_s}{\partial z}\right)^2 + \left(\frac{\partial v_s}{\partial z}\right)^2\right\}\right]dz + \frac{1}{2}\int_0^l P(z)\left(\frac{\partial w_b}{\partial z} + \frac{\partial w_b}{\partial z}\right)^2 dz \\
&+ \frac{1}{2}\int_0^l P(z)\left(\frac{\partial v_s}{\partial z} + \frac{\partial v_s}{\partial z}\right)^2 dz - \int_0^l p_w(z)(w_b + w_s)dz - \int_0^l p_v(z)(v_b + v_s)dz
\end{aligned} \tag{12.1}
$$

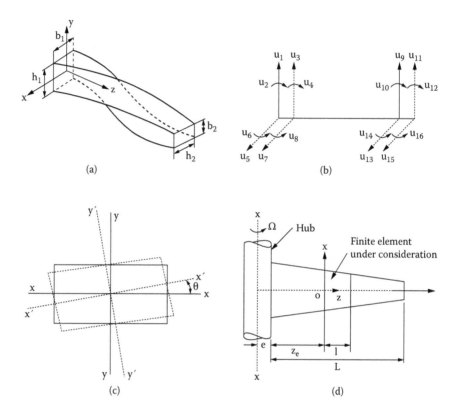

FIGURE 12.1
(a) An element of tapered and twisted beam. (b) Degrees of freedom of an element. (c) Angle of twist *y*. (d) Rotation of tapered beam. (From Kumar, S., et al., *Mechanical Systems and Signal Processing* 21(1):480–501, 2007. With permission.)

where

$$P(z) = \int_{e+z_e+z}^{L+e} m\Omega^2 \xi \, d\xi = \frac{\rho A \Omega^2}{2g} \left[(L+e)^2 - (e+z_e+z)^2 \right]$$

$$= \frac{\rho A \Omega^2}{g} \left[\left(eL + \frac{1}{2}L^2 - ez_e - \frac{1}{2}z_e^2 \right) - (e+z_e) z - \frac{1}{2}z^2 \right]$$

(12.2)

$$p_w(z) = \frac{\rho A \Omega^2}{g} (w_b + w_s)$$

(12.3)

$$p_v(z) = \frac{\rho A \Omega^2}{g} (v_b + v_s)$$

(12.4)

The kinetic energy T of the element, including the effect of shear deformation and rotary inertia, is given by

$$T = \int_0^l \left[\frac{\rho A}{2g} \left(\frac{\partial w_b}{\partial t} + \frac{\partial w_s}{\partial t} \right)^2 + \frac{\rho A}{2g} \left(\frac{\partial v_b}{\partial t} + \frac{\partial v_s}{\partial t} \right)^2 + \frac{\rho I_{yy}}{2g} \left(\frac{\partial^2 v_b}{\partial z \partial t} \right)^2 \right.$$

$$\left. + \frac{\rho I_{xy}}{g} \left(\frac{\partial^2 w_b}{\partial z \partial t} \right) \left(\frac{\partial^2 v_b}{\partial z \partial t} \right) + \frac{\rho I_{xx}}{2g} \left(\frac{\partial^2 w_b}{\partial z \partial t} \right)^2 \right] dz$$

(12.5)

The deformations can be discretized in terms of shape functions. Cubic polynomials are used for the out-of-plane bending w_b and in-plane bending v_b. Cubic polynomials are also used as shape functions for the shear deformations w_s and v_s. Using the energy expressions and the finite element discretization, the element level mass and stiffness matrices are calculated. After assembling the matrices and applying the cantilever boundary conditions we obtain

$$\left([K] - \omega^2 [M] \right) \Phi = 0$$

(12.6)

The stiffness and mass matrices are obtained by using expressions for the strain energy and kinetic energy. The details of the formulation can be found in [120], and the finite element model is validated by comparing with the results in [120].

12.1.2 Fatigue Damage Model

There are many models for predicting damage growth due to fatigue. Some LCF models proposed in the past have limited validity to particular cases [123, 124]. The physical meaning of the model parameters is also not clear. For engineering applications, models based on continuum damage mechanics appear useful. They do not require detailed models of crack growth using fracture mechanics, but capture the nonlinear nature of damage growth. They are based on the continuum damage variable D, which can be defined [131] as $D = 1 - E/E_0$.

In continuum damage mechanics, a damaged coupled potential is used as a starting point:

$$\Psi = \Psi_e \left(\varepsilon_{ij}^e, T, D, \pi \right)$$

(12.7)

For linear elasticity and isotropic damage, coupled damage constitutive equations are [132]

$$\sigma_{ij} = \rho \frac{\partial \Psi}{\partial \varepsilon_{ij}^e}$$

(12.8)

or

$$\varepsilon_{ij}^{e} = \frac{1+v}{E(1-D)}\sigma_{ij} - \frac{v}{E(1-D)}\sigma_{ij}\delta_{ij} \tag{12.9}$$

The damage strain energy release rate variable Y associated with D is defined by

$$Y = -\rho\frac{\partial\Psi}{\partial D} = -\frac{\sigma_{eq}^{2}R_{v}}{2E(1-D)^{2}} \tag{12.10}$$

where R_v is expressed for fatigue load as

$$R_{v} = \frac{2}{3}(1+v)+3(1-2v)\left(\frac{\sigma_{H}}{\sigma_{eq}}\right)^{2} \tag{12.11}$$

where σ_H is the hydrostatic stress defined by $\sigma_H = \sigma_{kk}/3$, σ_{eq} is the Von Mises equivalent stress defined by $\sigma_{eq} = \sqrt{3.S_{ij}S_{ij}/2}$, and S_{ij} is the stress deviator by $S_{ij} = \sigma_{ij} - \sigma_H\delta_{ij}$.

Assuming plastic deformation and microplastic deformation to cause damage and internal energy dissipation, the dissipate potential φ is

$$\varphi = \varphi_P(\sigma, R, D) + \varphi_D(Y, \dot{p}, \dot{\pi}, T, \varepsilon_e, D) + \varphi_\pi \tag{12.12}$$

where

$$\varphi_p = \frac{\sigma_{eq} - R}{1-D} - \sigma_Y \tag{12.13}$$

There is little information about microplastical dissipation potential φ_π, which is not considered here. Then the coupled damage constitutive equations and the dynamic damage evolution law can be derived from the plastic dissipated potential φ_p and the damage dissipated potential φ_D as follows:

$$\varepsilon_{ij} = \varepsilon_{ij}^{e} + \varepsilon_{ij}^{p} \tag{12.14}$$

$$\dot{\varepsilon}_{ij}^{p} = \lambda\frac{\partial\varphi_p}{\partial\sigma_{ij}} = \frac{3}{2}\frac{\lambda}{(1-D)}\frac{S_{ij}}{\sigma_{eq}} \tag{12.15}$$

$$\dot{p} = -\lambda\frac{\partial\varphi_p}{\partial R} = \frac{\lambda}{(1-D)}\left[\frac{2}{3}\dot{\varepsilon}_{ij}^{p}\dot{\varepsilon}_{ij}^{p}\right]^{1/2} \tag{12.16}$$

$$\dot{D} = -\lambda \frac{\partial \varphi_D}{\partial Y} \tag{12.17}$$

where λ is a nonnegative proportion factor, which can be obtained from the consistency condition, $\varphi_p = 0$.

Fatigue damage is mainly caused by accumulated plastic strain. According to the CDM theory, the LCF damage evolution law can be described by a suitable dissipation potential. At the basis of the damage potential function Equation (12.18), this is sufficient to model all the main properties within the hypothesis of isotropy damage.

$$\varphi = \frac{Y^2}{2S_0} \frac{\dot{p}}{(1-D)^{\alpha_0}} \tag{12.18}$$

The damage character of LCF is conceded and a dissipation potential φ is chosen as follows:

$$\varphi = \frac{Y^2}{2S_0} \frac{\Delta \dot{p}}{\left(1-\left(N/N_f\right)\right)^{(1-N_f\alpha)}} \tag{12.19}$$

The term $(1-N/N_f)$, other than $(1 - D)$, reflects the influence of accumulated plastic strain, and α is a parameter that describes the extent of accumulated damage; here it is the plastic strain increment per cycle, which can be determined from monotonic tensile and cyclic tensile stress-strain curves. Equation (12.19) can be written as

$$\dot{D} = -\frac{\partial \varphi}{\partial Y} = \left(-\frac{Y}{S_0}\right) \frac{\Delta \dot{p}}{\left(1-\left(N/N_f\right)\right)^{(1-N_f\alpha)}} \tag{12.20}$$

From Equations (12.10) and (12.20) one can get

$$\dot{D} = -\frac{\Delta \sigma_{eq}^2 R_v}{2ES_0(1-D)^2} \frac{\Delta \dot{p}}{\left(1-\left(N/N_f\right)\right)^{(1-N_f\alpha)}} \tag{12.21}$$

According to Lemaitre's hypothesis of strain equivalence, the cyclic stress-strain relationship coupled with damage should be written as follows:

$$\frac{\Delta \sigma_{eq}}{1-D} = K(\Delta p)^M \tag{12.22}$$

Equations (12.21) and (12.22) give the general constitutive equation for LCF damage:

$$\dot{D} = -\frac{K^2 R_v}{2ES_0} \frac{\Delta \dot{p}}{\left(1-\left(N/N_f\right)\right)^{(1-N_f\alpha)}} \tag{12.23}$$

In the case of the proportional loading per cycle, R_v can be considered constant with respect to time, and the damage during one cycle may be obtained through the integration of Equation (12.23):

$$\frac{\delta D}{\delta N} = \frac{K^2 R_v}{2ES_0} \frac{\Delta p^{2M+1}}{(2M+1)\left(1-\left(N/N_f\right)\right)^{1-N_f\alpha}} \tag{12.24}$$

Integrating Equation (12.24) with the initial conditions, $D|_{N=N_0}=D_0$, $D|_{N=N_f}=1$, gives

$$1-D_0 = \frac{K^2 R_v}{2ES_0} \frac{\Delta p^{2M+1}}{(2M+1)} \frac{1}{\alpha} \tag{12.25}$$

and

$$D-D_0 = \frac{K^2 R_v}{2ES_0} \frac{\Delta p^{2M+1}}{(2M+1)} \frac{1}{\alpha} \left[1-\left(1-\frac{N}{N_f}\right)^{N_f\alpha}\right] \tag{12.26}$$

Comparison of Equation (12.25) with Equation (12.26) gives the general LCF damage accumulation law:

$$D = 1-(1-D_0)\left(1-\frac{N}{N_f}\right)^{N_f\alpha} \tag{12.27}$$

There are three parameters in the above model: D_0, N_f, and $N_f\alpha$. These parameters can be determined from experiments. The damage model for steam turbine blade material 2Cr13 martensitic stainless steel is derived using experiments. In [133], a material testing machine was used for strain-controlled fatigue tests. There are several methods for measuring the damage variable D. For LCF, the best mechanisms include elasticity modulus followed by the cycle stress amplitude method. For measurements based on elastic stiffness, a specimen of the material needs to be machined to run mechanical tests. Here $D = 1 - (E/E_0)$ is used. This method needs accurate strain measurements. Typically, strain gauges are used and E is most accurately measured during unloading. Another approach that is used by [133] is called the cycle stress amplitude method. The one-dimensional law of cyclic plasticity at stabilization may be written as a power relationship between the amplitude of stress range $\Delta\sigma$ and the amplitude of strain range $\Delta\varepsilon$ at cycles:

$$\Delta\varepsilon = \left(\frac{\Delta\sigma}{K_c}\right)^M \tag{12.28}$$

The above relationship is for an undamaged material. K_c and M are material parameters. For damaged material, the relationship becomes

$$\Delta\varepsilon = \left[\frac{\Delta\sigma}{K_c(1-D)}\right]^M \tag{12.29}$$

For a cyclic test at a constant amplitude of strain $\Delta\varepsilon$, an initial stress being $\Delta\sigma_0$, the damage D may be assumed to be zero, and hence

$$\Delta\sigma_0 = K_c\Delta\varepsilon^{1/M} \tag{12.30}$$

and also from Equation (12.29),

$$\Delta\sigma = (1-D)K_c\Delta\varepsilon^{1/M} \tag{12.31}$$

Combining Equations (12.30) and (12.31) we obtain

$$D = 1 - \frac{\Delta\sigma}{\Delta\sigma_0} \tag{12.32}$$

The models developed for damage growth are as follows:

1. For low strain of ±0.35 per cycle the number of cycles (N_f) to produce fatigue crack resulting in failure was 6230 and the model was

$$D(N) = 1 - 0.906\left(1 - \frac{N}{6230}\right)^{0.058} \tag{12.33}$$

   where D is the damage level in the structure after N cycles. For $N = N_f = 0.9$, $D = 0.21$, with 623 cycles left to final failure.

2. For moderate strain of ±0.5 per cycle the number of cycles (N_f) to produce fatigue crack resulting in failure was 1950 and the model was

$$D(N) = 1 - 0.903\left(1 - \frac{N}{1950}\right)^{0.064} \tag{12.34}$$

   For $N/N_f = 0.9$, $D = 0.27$, with 195 cycles left for failure.

3. For high strain of ±0.70 per cycle the number of cycles (N_f) to produce fatigue crack resulting in failure was 844 and the model was

$$D(N) = 1 - 0.923\left(1 - \frac{N}{844}\right)^{0.113}$$ (12.35)

For $N/N_f = 0.9$, $D = 0.23$, with 84 cycles left for failure.

Note that the high-strain conditions where LCF happens occur only occasionally in the life of the actual machine. For example, the average design life of a steam turbine is about 30 years. Assuming the frequency of both cold start and sliding parameter stop is three times per year, the total number of cold starts and sliding parameter stops is 90 times in 30 years [126].

The damage curves are shown in Figure 12.2. The plots show that D varies quickly at the last stages of the whole cycling and slowly at the middle stage, from 10% to 80% of the total cycles, which is a characteristic of LCF damage. High strain leads to faster failure of the material. In general, the LCF damage curve can be divided into three stages. Stage 1 occurs up to $N/N_f = 0.1$, when damage value increases due to changes in the dislocation substructures. In stage 2, there is a slow increase in damage value up to $N/N_f = 0.8$. In stage 3, the damage increases more quickly to 1 due to the beginning of damage localization and formation of fatigue microcracks. From a condition monitoring viewpoint, it is useful to know when the blade has reached stage 3.

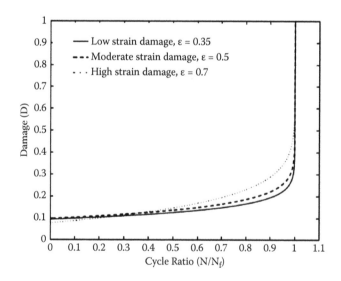

FIGURE 12.2
Damage variations with number of cycles for different strain load cases. (From Kumar, S., et al., *Mechanical Systems and Signal Processing* 21(1):480–501, 2007. With permission.)

The midpoint of stage 3 of $N/N_f = 0.9$ is therefore selected as the time where the life of the blade is almost over and the blade needs to be replaced.

12.1.3 Beam with Fatigue Damage

We want to know if fatigue damage can be detected in turbine blades before it becomes catastrophic by monitoring natural rotating frequencies. For the numerical results, the damage model discussed in the fatigue damage model is included in the finite element model in modeling of the turbine blade. The stiffness (E) of the beam and the fatigue damage growth, with number of cycles, is related with the expression [131] $D(N) = 1 - (E/E_0)$. Thus, Young's modulus for each case, i.e., for low-, moderate-, and high-strain models, can be written as follows:

For low-strain model:

$$E(N) = E_0 \left\{ 0.906 \left(1 - \frac{N}{6230} \right)^{0.058} \right\} \tag{12.36}$$

For moderate-strain model:

$$E(N) = E_0 \left\{ 0.903 \left(1 - \frac{N}{1950} \right)^{0.064} \right\} \tag{12.37}$$

For high-strain model:

$$E(N) = E_0 \left\{ 0.923 \left(1 - \frac{N}{844} \right)^{0.113} \right\} \tag{12.38}$$

The expression of Young's modulus for three cases (i.e., low, moderate, and high strains) is used in energy expressions to obtain stiffness and mass matrices for the damaged beam at any given point in time N.

12.2 Numerical Simulations

The baseline undamaged blade has a length of 0.254 m, depth at root of 0.000865 m, breath at root of 0.0173 m, twist of 45°, $\rho = 7800 \text{kg/m}^3$, and $E = 2.1\text{‰ } 1011 \text{ N/m}^2$. The beam is divided into 25 elements of equal length, resulting in elements of length equal to 4% of the beam length. The root, inboard, and outboard locations have five elements each. For numerical results, 20% length of the beam is damaged for each of three locations, i.e., considering stiffness reduction in 20% length of the beam in each of the root, inboard,

TABLE 12.1

Reduction in Stiffness with Number
of Cycles for High-Strain Condition

N/N_f	E/E_0
0	0.923
0.2	0.9
0.4	0.871
0.6	0.832
0.8	0.77
0.9	0.712
0.97	0.612
0.999	0.423
0.9999	0.326

Source: Kumar, S., et al., *Mechanical Systems and Signal Processing* 21(1):480–501, 2007. With permission.

and outboard locations. The stiffness reduction with the number of the cycles according to Equations (12.35) and (12.38) for the high-strain loading case is given in Table 12.1.

12.2.1 Finite Element Simulations

A finite element analysis of the beam considering reduction in stiffness at root, inboard, and outboard locations is carried out. It is found that for 25 elements the finite element solution converges even with considerable reduction in stiffness at the selected root, inboard, and outboard locations. Figure 12.3 shows the results of a convergence study for damage at root, inboard, and outboard locations as the number of finite elements increases to 25. In these cases, the damage value of $D = 0.99$ is considered at each level. The frequencies were normalized with the values obtained with 25 finite elements. A computer program is written for the calculation of the modal frequencies separately, considering stiffness reduction at root, inboard, and outboard locations and for different strain conditions.

Young's modulus values in Table 12.1 are used in energy expressions to get the mass and stiffness matrix. The eigenvalue problem is then solved to get the modal frequencies. The graphs in Figure 12.4 show the variation of frequencies with number of cycles for high-strain loading conditions. The data in the figures are normalized with respect to a baseline undamaged blade. For the root location in Figure 12.4, the maximum change is in the fourth mode frequency. For the inboard location, the third mode changes most, and for the outboard location, the higher mode changes most.

Figure 12.4 also shows results for the low-strain case, and it is evident from the figure that damage at the root location causes maximum frequency

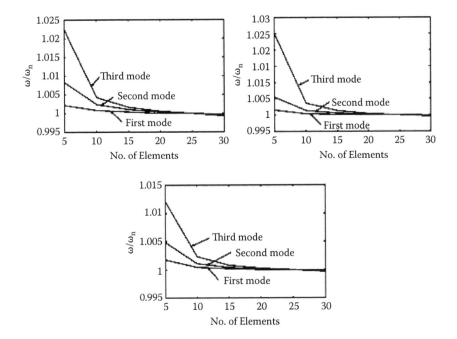

FIGURE 12.3
Convergence trend with number of elements for different locations of the blade. (From Kumar, S., et al., *Mechanical Systems and Signal Processing* 21(1):480–501, 2007. With permission.)

change in the first mode, while damage at the inboard and outboard locations affects the third and sixth modes, respectively. The moderate strain results are also shown in Figure 12.4. Here the root and inboard damage locations affect the third mode, and the outboard damage location affects the higher modes. These graphs give the deterioration curves for frequency with respect to damage and show that rotating frequencies can be used as a virtual indicator of the damage variable D.

Tables 12.2–12.4 show the actual rotating frequencies in Hertz as the damage growth progresses. These results are for the high-strain case and are based on the underlying data in Figure 12.4. They show that measurable changes in the frequencies occur due to LCF damage. Consider the location $N/N_f = 0.9$, which is slightly before complete failure of the material. Table 12.5 shows the reduction in frequencies at these points in Hertz. The table shows that damage growth at the root location is easier to detect since it affects the lower modes. Outboard damage affects the higher modes to a greater extent.

12.2.2 Damage Detection

The reduction in frequency $\Delta\omega$ can be considered to be a health residual, which can be tracked for condition monitoring of the turbine blade. Ideally,

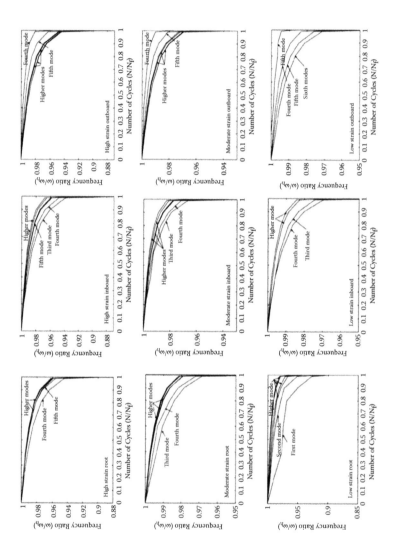

FIGURE 12.4
Frequency variations with number of cycles for stiffness reduction at different locations and strain cases of the blade. (From Kumar, S., et al., *Mechanical Systems and Signal Processing* 21(1):480–501, 2007. With permission.)

TABLE 12.2

Modal Frequencies (Hz) for Reduction in Stiffness at Root Location of the Beam for High-Strain Condition

N/N_f	0	0.2	0.4	0.6	0.8	0.9	0.97	0.999	0.9999
Mode 1	165.99	165.16	164.35	160.34	157.44	153.43	151.81	136.34	120.84
Mode 2	298.23	296.4	294.319	292.03	285.36	280.43	268.34	242.06	222.33
Mode 3	635.75	633.44	630.95	625.83	618.3	610.84	598.08	568.04	549.72
Mode 4	1061	1057.34	1052.4	1045.8	1034.23	1022.99	1002.4	957.46	929.48
Mode 5	1544.12	1540.18	1535.22	1527.69	1515.62	1503.67	1481.66	1428.76	1391.09
Mode 6	2403.38	2397.82	2390.81	2380.28	2362.85	2345.16	2310.72	2221.29	2151.31
Mode 7	2955.1	2948.42	2939.81	2927.39	2906.61	2885.31	2843.53	2733.22	2646.56
Mode 8	4146.55	4136.28	4122.74	4103.51	4070.1	4035.28	3965.2	3783.75	3654.61
Mode 9	4996.86	4985.54	4970.65	4949.48	4912.88	4874.68	4797.22	4586.11	4421.04
Mode 10	6207.42	6190.74	6169.01	6138.42	6086.5	6033.97	5933.46	5702.74	5556.83

Source: Kumar, S., et al., *Mechanical Systems and Signal Processing* 21(1):480–501, 2007. With permission.

TABLE 12.3

Modal Frequencies (Hz) for Reduction in Stiffness at Inboard Location of the Beam for High-Strain Condition

N/N_f	0	0.2	0.4	0.6	0.8	0.9	0.97	0.999	0.9999
Mode 1	171.42	169.67	169.07	168.65	165.85	164.54	163.69	155.97	145.54
Mode 2	300.38	296.89	299.78	297.23	294.56	290.50	287.83	269.31	255.32
Mode 3	637.40	635.54	631.90	628.15	620.45	612.45	598.34	562.40	536.74
Mode 4	1063.11	1060.04	1055.82	1049.87	1039.34	1028.62	1008.21	955.79	919.25
Mode 5	1548.80	1546.16	1542.16	1536.96	1527.48	1517.86	1498.82	1448.92	1410.14
Mode 6	2405.46	2400.38	2393.76	2384.55	2368.69	2352.65	2321.48	2243.40	2186.17
Mode 7	2953.70	2946.47	2937.04	2923.93	2901.42	2878.73	2834.77	2726.88	2649.36
Mode 8	4146.70	4136.44	4123.05	4104.23	4072.09	4039.24	3974.47	3808.27	3684.87
Mode 9	4992.09	4979.55	4963.22	4940.33	4901.29	4861.51	4783.67	4586.87	4440.85
Mode 10	6212.24	6196.72	6176.45	6147.90	6099.08	6049.07	5950.95	5706.31	5534.49

Source: Kumar, S., et al., *Mechanical Systems and Signal Processing* 21(1):480–501, 2007. With permission.

TABLE 12.4

Modal Frequencies (Hz) for Reduction in Stiffness at Outboard Location of the Beam for High-Strain Condition

N/N_f	0	0.2	0.4	0.6	0.8	0.9	0.97	0.999	0.9999
Mode 1	177.89	176.96	172.20	171.42	169.76	169.22	168.30	166.78	166.10
Mode 2	304.94	304.66	304.01	304.01	302.52	302.47	302.20	301.50	300.07
Mode 3	643.57	643.01	641.91	641.91	641.22	638.11	634.75	626.66	617.21
Mode 4	1070.88	1069.79	1068.76	1068.76	1062.68	1059.22	1050.99	1024.61	999.16
Mode 5	1549.26	1545.91	1542.33	1542.33	1527.05	1515.47	1492.51	1424.29	1367.43
Mode 6	2401.17	2394.42	2385.67	2385.67	2350.85	2327.21	22,279.32	2151.52	2056.57
Mode 7	2950.25	2941.48	2930.12	2930.12	2885.85	2855.87	2796.34	2641.55	2532.72
Mode 8	4134.33	4120.42	4102.24	4102.24	4034.36	3991.72	3911.27	3728.72	3610.68
Mode 9	4979.44	4961.59	4939.70	4939.70	4856.66	4803.74	4702.62	4470.47	4323.60
Mode 10	6203.51	6185.76	6162.79	6130.82	6077.88	6024.68	5922.72	5663.77	5454.94

Source: Kumar, S., et al., *Mechanical Systems and Signal Processing* 21(1):480–501, 2007. With permission.

TABLE 12.5

Reduction in Frequencies (Hz) at $N/N_f = 0.9$

Mode	High Strain			Moderate Strain			Low Strain		
	Root	Inboard	Outboard	Root	Inboard	Outboard	Root	Inboard	Outboard
1	12.56	6.88	8.67	0.85	0.15	0.75	8.06	4.26	0.24
2	17.8	9.88	2.47	8	6.95	1.26	13.4	12.5	2.28
3	24.89	24.95	5.46	13.65	13.25	3.04	19.8	17.58	5.74
4	38.01	34.69	11.66	19.34	19.97	6.56	20.71	15.15	16.49
5	40.45	31.13	33.79	21.04	17.2	18.3	29.86	27.1	36.92
6	58.22	52.81	73.96	29.8	30.18	41.23	35.56	38.37	47.23
7	69.79	74.99	94.38	35.6	42.56	52.78	56.09	54.87	73.76
8	111.27	107.46	142.61	55.96	60.89	81.52	61.62	66.77	89.46
9	122.18	130.59	174.7	61.69	74.05	99.43	88.95	83.08	92.4
10	173.45	162.67	178.89	88.86	92.33	102.22	106.18	103.61	124.6

Source: Kumar, S., et al., *Mechanical Systems and Signal Processing* 21(1):480–501, 2007. With permission.

the residuals should only be affected by faults. However, even for an undamaged system, the presence of disturbance noise and modeling errors causes residuals to become nonzero and interferes with the detection of faults.

The two main sources of errors in this problem are due to finite element modeling and the presence of measurement noise. Modern finite element methods are quite accurate. For example, Lawson and Ivey [128] compared measured and finite element simulated frequencies of compressor rotor blades. For the first three modes, the measured frequencies were 244, 736, and 1471 Hz and the simulated frequencies were 243, 740, and 1486 Hz, showing a discrepancy of 0.4, 0.5, and 1.0%. Strain gauge signals were used for the measurements. So it is possible that some of the difference can be attributed to measurement noise and the rest to modeling errors. Finite element models can be further improved using model updating methods [134] to match with the experimental results, and the errors due to modeling can be minimized.

The second source of error is the presence of measurement noise. To some extent, processing the measurement by signal processing methods can reduce noise, as we have seen in the Chapters 2–7. Another way to make fault detection more robust is to establish thresholds on the residuals instead of just checking for nonzero values. A key problem in fault detection is the establishment of thresholds on the residuals that can be used to signal when damage has become sufficiently large to be dangerous. For example, for a scalar residual r(N) a threshold T can be established as:

If $r(N) < T$, then no fault.

If $r(N) \geq T$, then fault.

A schematic representation of a residual generator is shown in Figure 12.5. Such a fault residual can be obtained from experiments, numerical simulations, or in-service data. Typically, a threshold can be developed from

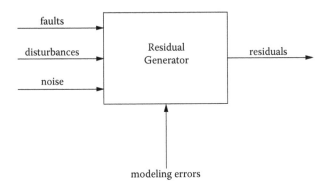

FIGURE 12.5
Schematic representation of a residual generator. (From Kumar, S., et al., *Mechanical Systems and Signal Processing* 21(1):480–501, 2007. With permission.)

numerical simulation and then refined based on in-service performance. Table 12.5 can be used to develop frequency thresholds on the health residuals $\Delta\omega = \omega^n - \omega^d$.

Note that for an ideal undamaged blade, $\Delta\omega = 0$. However, a turbine blade can take considerable damage before final failure. Therefore, the $\Delta\omega$ variations in Table 12.5 can be used to select a somewhat high threshold based on $N_f = 0.9$, where a possible end of life of the blade can be triggered. Considering the high-strain case, one can select the minimum $\Delta\omega$ between the root, inboard, and outboard locations as the threshold. This will lead to many false alarms but minimize missed alarms. On the other hand, if the maximum $\Delta\omega$ is selected, it will minimize false alarms and lead to more missed alarms. In general, threshold selection involves a trade-off between missed alarms and false alarms [135]. Furthermore, the presence of noise in the data can be addressed by slightly increasing the thresholds. Table 12.6 shows the thresholds based on the maximum criteria after being increased by 5%, which is a conservative design and minimizes false alarms.

These residuals can then be used to develop a fault detection system as shown in Figure 12.6. Here each residual is tested separately against an individual threshold. A simple rule is that if any frequency threshold is exceeded, an alarm is indicated. Thus, if any of the outputs from the test box in Figure 12.6 is 1, an alarm is indicated by the diagnostic system.

For ideal data, the maximum thresholds in Table 12.6 will give zero false alarms. The effect of increasing noise on the success rate is shown in Figure 12.7. Here noisy data are generated by adding noise to the ideal signals in Table 12.5 using

$$\Delta\omega_i^{(noisy)} = \Delta\omega_i\left(1+(\alpha/100)r\right) \tag{12.39}$$

TABLE 12.6

Frequency Thresholds for Damage Detection

Mode	High Strain	Moderate Strain	Low Strain
1	13.19	0.89	8.46
2	18.69	8.4	14.07
3	26.19	14.33	20.79
4	39.91	20.31	21.74
5	42.47	22.09	38.76
6	77.66	43.29	49.59
7	99.09	55.42	77.45
8	149.74	85.59	93.87
9	183.43	104.4	97.02
10	187.83	107.33	130.83

Source: Kumar, S., et al., *Mechanical Systems and Signal Processing* 21(1):480–501, 2007. With permission.

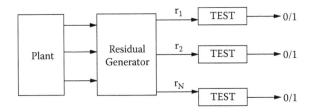

FIGURE 12.6
A schematic representation of a fault detection system. (From Kumar, S., et al., *Mechanical Systems and Signal Processing* 21(1):480–501, 2007. With permission.)

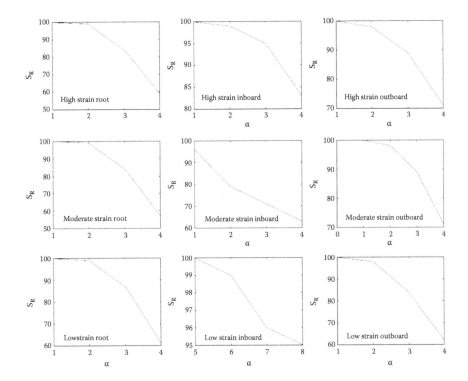

FIGURE 12.7
Effect of increasing percent noise in data on damage detection success rate. (From Kumar, S., et al., *Mechanical Systems and Signal Processing* 21(1):480–501, 2007. With permission.)

where α is a measure of noise and r is a random number between −1 and 1 from a normal distribution. Here $\alpha = 1$ corresponds to 1% noise in the measurement.

It is clear from Figure 12.7 that the detection algorithm has a very high accuracy with about 1–2% noise level in the data. According to Friswell and Jet [136], frequency can be measured to an accuracy up to 0.1%, so it can be said that the damage detection scheme based on maximum thresholds is robust.

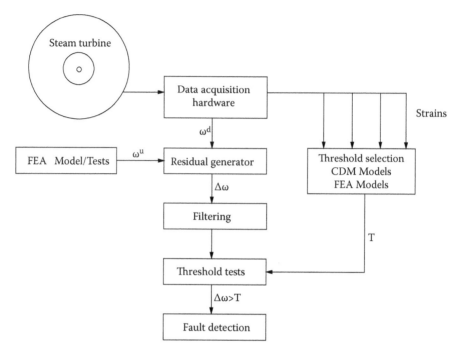

FIGURE 12.8
A schematic representation of a fault detection system. (From Kumar, S., et al., *Mechanical Systems and Signal Processing* 21(1):480–501, 2007. With permission.)

A schematic representation of the damage detection system is shown in Figure 12.8. Rotating frequencies are measured for the turbine blade and compared with undamaged results to obtain a frequency residual $\Delta\omega$. The residual is then low-pass filtered to remove noise. Appropriate thresholds are calculated for each frequency based on the strain level, CDM models, and finite element analysis. The residual is then threshold tested to determine if damage has occurred.

Vibration monitoring systems are very important for gas turbine diagnostics. In the introduction, a brief outline of these systems for the complete gas turbine was given, largely adapted from [137–139]. The current chapter has shown an application to turbine blades.

12.3 Summary

The chapter has shown that frequencies can be used to detect damage in a turbine blade just before it becomes catastrophic. Though we have considered a steam turbine blade, the approach is also applicable to gas turbine blades.

However, it should be noted that for any identification procedure, the error between the ideal and the estimation frequency also exists and is unavoidable. For example, the maximal reduction in percentage of natural frequency at $N / N_f = 0.9$ occurs for mode 1 and is about 7.57%. Other frequency reductions are about 4%. However, for modal parameter identification, an accuracy of about 7.57% is practically not easy to achieve using either frequency domain or time domain methods. Nevertheless, the accuracy of frequency measurement continues to improve rapidly based on newer sensors and signal processing methods. Using smart sensors leads to much less noise than strain gauges, etc., and therefore the identification of frequencies from the vibration data can be more accurate. Recent studies have looked at online estimation of frequencies. Oberholster and Heyns [140] developed a methodology for the online condition monitoring of axial-flow fan blades by using mode shapes and frequencies that were extracted from online blade vibration and strain signals. Further, filtering methods based on median and wavelet approaches can make the process of frequency extraction from raw data more accurate. In addition, though frequencies are not sensitive to small damage, this can be viewed as an advantage, as most real structures are designed to take considerable damage before failure [141]. The suggested method in this chapter therefore depends on the availability of accurate sensors and signal processing methods.

Though the current chapter has used frequencies as the damage indicators, CDM models with finite element simulations can also be used to track changes in blade response, strains, acceleration, and other measurable variables for the diagnostic applications. The effect of LCF damage on the rotating frequency of a turbine blade is studied. The turbine blade is modeled using a rotating Timoshenko beam with taper and twist and the frequencies obtained using finite element analysis. A damage model derived using continuum damage mechanics and identified from experimental data is used to accurately capture the nonlinear nature of LCF damage. It is found that LCF causes sufficient material degradation, resulting in stiffness loss as the damage growth progresses. The change in rotating frequency can be used as an indicator to track damage growth. Since turbine blades are capable of sustaining considerable accumulated damage before final failure, the simulated deterioration curves relating frequencies to damage are used to determine thresholds on the frequencies at the point where 90% of blade life is consumed. By placing suitable thresholds for the residual frequency, it is possible to detect the onset of the final stage of damage in the structure before final failure.

The chapter marks a considerable advance over most works on damage detection, which do not address the issue of damage growth and focus on the identification and detection of damage at only one time point. Coupling of vibration-based methods with performance-based methods such as those discussed in earlier chapters is a good approach to increase the accuracy and reliability of gas turbine diagnostics.

References

1. Sieros, G., Stamasis, K., and Mathioudakis, K. 1997. Jet Engine Component Maps for Performance Modeling and Diagnosis. *Journal of Propulsion and Power* 13(5):665–674.
2. Pinelli, M., and Spina, P.R. 2002. Gas Path Field Performance Determination: Sources of Uncertainties. *ASME Journal of Engineering for Gas Turbine and Power* 124(1):155–160.
3. Li, Y.G. 2002. Performance Analysis Based Gas Turbine Diagnostics: A Review. *Proceedings of the I MECH E Part A Journal of Power and Energy* 216(5):363–377.
4. Fasching, W.A., and Stricklin, R. 1982. *CF6 Jet Engine Diagnostics Program: Final Report*. NASA/CR-165582.
5. Mathioudakis, K., Kamboukos, Ph., and Stamasis, A. 2002. Turbofan Performance Deterioration Tracking Using Non-Linear Models and Optimization Techniques. *ASME Journal of Turbomachinery* 124(4).
6. DePold, H., and Gass, F.D. 1999. The Application of Expert Systems and Neural Networks to Gas Turbine Prognostics and Diagnostics. *ASME Journal of Engineering for Gas Turbine and Power* 121(4):607–612.
7. Doel, D.L. 1994. TEMPER-A Gas-Path Analysis Tool for Commercial Jet Engines. *ASME Journal of Engineering for Gas Turbine and Power* 116(1):82–89.
8. Doel, D.L. 2002. Interpretation of Weighted Least Squares Gas Path Analysis Results. Presented at Proceedings of the 47th ASME Gas Turbine and Aeroengine Technical Conference, Amsterdam, June 3–6.
9. Urban, L.A., and Volponi, A.J. 1992. Mathematical Methods of Relative Engine Performance Diagnostics. SAE Technical Paper 922048. *SAE Journal of Aerospace* 101.
10. Zedda, M., and Singh, R. 2000. Neural Network Based Sensor Validation for Gas Turbine Test Bed Analysis. *Proceedings of I MECH E Part I Journal of Systems and Control in Engineering* 215(1):47–56.
11. Lu, P.J., Hsu, T.C., Zhang, M.C., and Zhang, J. 2001. An Evaluation of Engine Fault Diagnostics Using Artificial Neural Networks. *ASME Journal of Engineering for Gas Turbine and Power* 123(2):240–246.
12. Volponi, A.J., Depold, H., Ganguli, R., and Daguang, C. 2000. The Use of Kalman Filter and Neural Network Methodologies in Gas Turbine Performance Diagnostics: A Comparative Study. *ASME Journal of Engineering for Gas Turbine and Power* 125(4):917–924.
13. Ganguli, R. 2002. Fuzzy Logic Intelligent System for Gas Turbine Module and System Fault Isolation. *Journal of Propulsion and Power* 18(2):440–447.
14. Romessis, C., Stamatis, A., and Mathioudakis, A.K. 2001. *Setting up a Belief Network for Turbofan Diagnosis with the Aid of an Engine Performance Model*. ISABE Paper 1032.
15. Sugiyama, N. 2000. System Identification of Jet Engines. *ASME Journal of Engineering for Gas Turbine and Power* 122(1):19–26.

16. Kalluri, S., and Arce, G.R. 1998. Adaptive Weighted Myriad Filter Algorithms for Robust Signal Processing in α-Stable Noise Environment. *IEEE Transactions on Signal Processing* 46(2):322–334.

17. Windyga, P.S. 2001. Fast Impulsive Noise Removal. *IEEE Transactions on Image Processing* 10(2):173–179.

18. Schroeder, W., Martin, K., and Lorensen, B. 1998. *The Visualization Toolkit*. Prentice-Hall: New Jersey, 432–435.

19. Yin, L., Yang, M., Gabbouj, M., and Neunou, Y. 1996. Weighted Median Filters: A Tutorial. *IEEE Transactions on Circuits and Systems* 40(1):147–192.

20. Heinonen, P., and Neuvo, Y. 1987. FIR-Median Hybrid Filter. *IEEE Transactions on Acoustics, Speech and Signal Processing* 35:832–838.

21. Senel, H.G., Peters II, A.R., and Dawant, B. 2002. Topological Median Filters. *IEEE Transactions on Image Processing* 11(2):89–104.

22. Sun, T., Gabbouj, M., and Neunou, Y. 1994. Center Weighted Median Filters: Some Properties and Their Applications in Image Processing. *Signal Processing* 35(3):213–229.

23. Chen, T., Ma, K.K., and Chen, L.H. 1999. Tri-State Median Filter for Image Denoising. *IEEE Transactions on Image Processing* 8(12):1834–1838.

24. Chen, T., and Wu, R.H. 2001. Adaptive Impulse Detection Using Center Weighted Median Filters. *IEEE Signal Processing Letters* 8(1):1–3.

25. Ganguli, R. 2002. Data Rectification and Detection of Trend Shifts in Jet Engine Gas Path Measurements Using Median Filters and Fuzzy Logic. *ASME Transactions: Journal of Engineering in Gas Turbine and Power* 124(4):809–816.

26. Nounou, M.N., and Bakshi, B.R. 1999. On-Line Multiscale Filtering of Random and Gross Errors without Process Models. *AIChE Journal* 45(5):1041–1058.

27. Manders, E.J., Biswas, G., Mosterman, P.J., Barford, L., Ran, V., and Bennet, J. 2000. Signal Interpretation for Monitoring and Diagnosis, a Cooling System Testbed. *Proceedings of the 16th IEEE Instrumentation and Measurement Technology Conference IMTC* 99:498–503.

28. Ogaji, O., Rizos, C., and Wang, J. 2001. A Dynamic GPS System for On-Line Structural Monitoring. In *International Symposium on Kinematic Systems in Geodesy, Geomatices and Navigation (KIS)*, Barff, Canada, June 5–8, 290–297.

29. Kramer, M.A. 1991. Non-Linear Principal Component Analysis Using Autoassociative Networks. *AIChE Journal* 37(2):233–243.

30. Kramer, M.A. 1992. Autoassociative Neural Networks. *Computers and Chemical Engineering* 16(4):313–328.

31. Lu, P.J., and Hsu, T.C. 2002. Application of Autoassociative Neural Network on Gas Path Sensor Data Validation. *AIAA Journal of Propulsion and Power* 18(4):879–888.

32. Staszewski, W.J. 2002. Intelligent Signal Processing for Damage Detection in Composite Materials. *Composite Science and Technology* 62:941–950.

33. Staszewski, W.J. 2000. Advanced Data Preprocessing for Damage Identification Based on Pattern Recognition. *International Journal as Systems Science* 31(11):1381–1396.

34. Urban, L.A. 1975. Parameter Selection for Multiple Fault Diagnostics of Gas Turbine Engines. *Journal of Engineering for Power* 225–230.

35. Wendt, P.D., Coyle, E.J., and Gallagher Jr., N.C. 1986. Some Convergence Properties of Median Filters. *IEEE Transactions on Circuits and Systems* 33:276–286.

36. D.L. Doel. 1990. Role for Expert Systems in Commercial Gas Turbine Engine Monitoring. ASME Paper GT37411.

37. Ganguli, R. 2003. Jet Engine Gas Path Measurement Filtering Using Center Weighted Idempotent Median Filters. *Journal of Propulsion and Power* 19(5):930–937.
38. Arce, G.R., and Parades, J.L. 2000. Recursive Weighted Median Filter Admitting Negative Weights and Their Optimization. *IEEE Transactions on Signal Processing* 48(3):768–779.
39. Haavista, P., Juhola, J., and Neunou, Y. 1991. Median Based Idempotent Filters. *Journal of Circuits Systems Computing* 1(2):125–148.
40. Volponi, A.J. 1999. Gas Turbine Parameter Corrections. *ASME Journal of Engineering for Gas Turbine and Power* 121(4):613–621.
41. Verma, R., and Ganguli, R. 2005. Denoising Jet Engine Gas Path Measurements Using Nonlinear Filters. *IEEE/ASME Transactions on Mechatronics* 10(4):461–464.
42. Ramponi, G. 1996. The Rational Filter for Image Smoothing. *IEEE Signal Processing Letters* 3(3).
43. Ganguli, R. 2002. Noise and Outlier Removal from Jet Engine Health Signals Using Weighted FIR Median Hybrid Filters. *Mechanical Systems and Signal Processing* 16(6):967–978.
44. Neerjarvi, J., Varri, A., Fotopoulos, S., and Neuvo, Y. 1993. Weighted FMH Filters. *Signal Processing* 31:181–190.
45. Doel, D.L. 2003. Interpretation of Weighted-Least-Squares Gas Path Analysis Results. *Journal of Engineering for Gas Turbine and Power* 125(3):624–633.
46. Li, Y.G. 2003. A Gas Turbine Diagnostic Approach with Transient Measurements. *Journal of Power and Energy* 217(2):169–177.
47. Luppold, R., Roman, J.R., Gallops, G.W., and Kerr, L.J. 1989. *Estimating In-Flight Engine Performance Variations Using Kalman Filter Concepts*. AIAA-89-2584.
48. Merrington, G.L. 1988. *Identification of Dynamic Characteristics for Fault Isolation Purposes in a Gas Turbine Using Closed-Loop Measurements*. Engine Condition Monitoring—Technology and Experience. AGARD-CP-448.
49. Merrington, G.L. 1994. Fault Diagnosis in Gas Turbines Using a Model-Based Technique. *Journal of Engineering for Gas Turbine and Power* 116:374–380.
50. Gonzalez, J.G., and Arce, G.R. 2002. Statistically-Efficient Filtering in Impulsive Environments: Weighted Myriad Filters. *EURASIP Journal of Applied Signal Processing* 1:4–20.
51. Gonzalez, J.G., and Arce, G.R. 2001. Optimality for the Myriad Filter in Practical Impulsive-Noise Environments. *IEEE Transactions on Signal Processing* 49(2):438–441.
52. Kalluri, S., and Arce, G.R. 2000. Fast Algorithm for Weighted Myriad Computation by Fixed Point Search. *IEEE Transactions on Signal Processing* 48(1):159–171.
53. Surender, V.P., and Ganguli, R. 2005. Adaptive Myriad Filter for Improved Gas Turbine Condition Monitoring Using Transient Data. *Journal of Engineering for Gas Turbine and Power* 127(2):329–339.
54. Ogaji, S.O.T., Li, Y.G., Sampath, S., and Singh, R. (2003). *Gas Path Fault Diagnosis of a Turbofan Engine from Transient Data Using Artificial Neural Networks*. ASME Paper GT2003-38423.
55. Ganguli, R., and Dan, B. 2004. Trend Shift Detection in Jet Engine Gas Path Measurement Using Cascaded Recursive Median Filter with Gradient and Laplacian Edge Detector. *Journal of Engineering for Gas Turbine and Power* 126(1):55–61.
56. Gonzales, R.C., and Woods, R.E. 2002. *Digital Image Processing*. Reading, MA: Addison Wesley.

57. Canny, J. 1986. A Computational Approach to Edge Detection. *IEEE Transactions on Pattern Analysis and Machine Intelligence* 8(6):679–698.
58. Marr, D., and Hildreth, E. 1980. Theory of Edge Detection. *Proceedings of Royal Society Series B, Biological Sciences* 207(1167):187–217.
59. Chou, P.C., and Bennamoun, M. 2000. Accurate Localization of Edges in Noisy Volume Images. Presented at Proceedings of the IEEE International Conference on Pattern Recognition, Piscataway, NJ.
60. Mao, M., and Gan, Z.J. 1993. Statistical Analysis for the Convergence Rate of Signals to Median Filter Roots. *IEEE Transactions on Signal Processing* 41:2499–2502.
61. Arce, G., and Gallagher, N.C. 1988. Stochastic Analysis for the Recursive Median Filter Process. *IEEE Transactions on Information Theory* 34(4):669–679.
62. Chen, T., and Wu, H.R. 2001. Recursive LMS L-Filters for Noise Removal in Images. *Signal Processing Letters* 8(2):36–38.
63. Shmulevich, I., and Coyle, E.J. 1997. The Use of Recursive Median Filters for Establishing the Tonal Context of Music. In *IEEE Workshop on Nonlinear Signal and Image Processing*, MI26.
64. Bangham, J.A. 1993. Properties of a Series of Nested Median Filters, Namely the Data Sieve. *IEEE Transactions on Signal Processing* 41:31–42.
65. Fitch, J.P., Coyle, E.J., and Gallagher, N.C. 1984. Median Filtering by Threshold Decomposition. *IEEE Transactions on Acoustics, Speech and Signal Processing* 1184–1188.
66. Richards, S.D. 1990. LSI Median Filters. *IEEE Transactions on Acoustics, Speech and Signal Processing* 38:145–153.
67. Alliney, S. 1996. Recursive Median Filters of Increasing Order: A Variational Approach. *IEEE Transactions on Signal Processing* 44(6):1346–1354.
68. Bangham, J.A., Ling, P., and Young, R. 1996. Multiscale Recursive Medians, Scale-Space and Transforms with Applications to Image Processing. *IEEE Transactions on Image Processing* 5(6):1043–1048.
69. Yli-Harja, O., Bangham, J.A., Harvey, R., and Aldridge, R. 1999. Correlation Properties of Cascaded Recursive Median Filters. In *IEEE Workshop on Nonlinear Signal and Image Processing*, Antalya, Turkey. Piscataway, NJ: IEEE, 491–495.
70. Yli-Harja, O., Koivisto, P., Bangham, J.A., Cawley, G., Harvey, R., and Schmulevich, I. 2001. Simplified Implementation of the Recursive Median Sieve. *Signal Processing* 81:1565–1570.
71. Yoshida, I. 2002. Health Monitoring Algorithm by Monte Carlo Filter Based on Non-Gaussian Noise. *Journal of Natural Disaster Science* 24(2):101–107G.
72. Uday, P., and Ganguli, R. 2010. Jet Engine Health Signal Denoising Using Optimally Weighted Recursive Median Filters. *Journal of Engineering for Gas Turbines and Power* 132(4).
73. Roy, N., and Ganguli, R. 2006. Filter Design Using Radial Basis Function Neural Network and Genetic Algorithm for Improved Operational Health Monitoring. *Applied Soft Computing* 6(2):154–169.
74. Guruprakash, V.N., and Ganguli, R. 2011. Three and Seven Point Optimally Weighted Recursive Median Filters for Gas Turbine Diagnostics. *ASME Journal of Engineering in Gas Turbine and Power* 133(10): article 104502.
75. Nelwamondo, F.V., and Marwala, T. 2008. Techniques for Handling Missing Data: Applications to Online Condition Monitoring. *International Journal of Innovative Computing Information and Control* 4(6):1507–1526.

76. Volponi, A.J. 2003. *Foundations of Gas Path Analysis I and II, Gas Turbine Condition Monitoring and Fault Diagnosis.* Von Karman Institute for Fluid Dynamics, Lecture Series 2003–01.

77. Urban, L.A. 1972. *Gas Path Analysis Applied to Turbine Engine Conditioning Monitoring.* AIAA/SAE Paper 72:1082.

78. Doel, D.L. 1992. Assessment of Weighted-Least-Squares-Based Gas Path Analysis. *ASME Journal of Engineering for Gas Turbine and Power* 116(2):366–373.

79. Merrington, G.L. 1993. *Fault Diagnosis in Gas Turbines Using a Model Based Technique.* ASME Paper 93-GT-13. Bellingham, WA: Society for Optical Engineering.

80. Winston, H., et al. 1991. *Integrating Numeric and Symbolic Processing for Gas Path Maintenance.* AIAA Paper 91-0501.

81. Gallops, G.W., et al. 1992. *In-Flight Performance Diagnostic Capability of an Adaptive Engine Model.* AIAA Paper 92:37–46.

82. Kerr, L.J., et al. 1999. *Real-Time Estimation of Gas Turbine Engine Damage Using a Control Based Kalman Filter Algorithm.* ASME Paper 91-GT-216.

83. Volponi, A.J. 1994. *Sensor Error Compensation in Engine Performance Diagnostics.* ASME Paper 94-GT-058.

84. McDuff, R.J., and Simpson, P.K. 1990. An Investigation of Neural Network for F-16 Fault Diagnosis. In *Proceedings of the SPIE Technical Symposium on Intelligent Information Systems.* Bellingham, WA: International Society for Optical Engineering.

85. Guo, Z., and Uhhrig, R.E. 1992. Using Modular Neural Networks to Monitor Accident Conditions in Nuclear Power Plants. In *Proceedings of the SPIE Technical Symposium on Intelligent Information Systems.* Bellingham, WA: International Society for Optical Engineering.

86. Wantanabe, K., et al. 1989. Incipient Fault Diagnosis of Chemical Processes via Artificial Neural Network. *AICHE Journal* 35(11).

87. Holmstrom, L., and Koistinen, P. 1992. Using Additive Noise in Back Propagation Training. *IEEE Transactions on Neural Networks* 3(1):24–38.

88. Haykin, S. 1994. *Neural Networks—A Comprehensive Foundation.* New York: Macmillan.

89. Dasarathy, B.V. 1991. *Nearest Neighbor (NN) Norms: NN Pattern Classification Techniques.* New York: IEEE Computer Society Press.

90. Kohonen, T. 1990. The Self-Organizing Map. In *Proceedings of IEEE*, New York.

91. Kirkpatrick, S. 1984. Optimization by Simulated Annealing: Quantitative Studies. *Journal of Statistical Physics* 34(5–6):975–986.

92. Ackley, D.H., Hinton, G.E., and Sejnowski, T.J. 1985. A Learning Algorithm for Boltzmann Machines. *Cognitive Science* 9:147–169.

93. Broomhead, D.S., and Lowe, D. 1988. *Multivariate Function Interpolation and Adaptive Networks. Complex System.* 2. Champaign, IL: Complex Systems Publications.

94. Guo, T.H., and Saus, J. 1996. Sensor Validation for Turbofan Engines Using an Autoassociative Neural Network. Presented at AIAA Guidance Navigation and Control Conference.

95. Sinha, N.K., and Gupta, M.M. 2000. *Soft Computing and Intelligent Systems: Theory and Applications.* San Diego: Academic Press.

96. Hornik, K., Stinchcombe, M., and White, H. 1989. Multilayer Feed Forward Networks Are Uniform Approximators. *Neural Networks* 2(3):366–369.

97. Luger, G.F., and Stubblefield, W.A. 1998. *Artificial Intelligence—Structures and Strategies for Complex Problem Solving.* Reading, MA: Addison Wesley Longman.

98. Yen, J., Laqngani, R., and Zadeh, L.A. (eds.). 1995. *Industrial Applications of Fuzzy Logic and Intelligent Systems.* New York: IEEE Press.

99. Hong, X.L., and Chen, P.C.L. 2000. The Equivalence between Fuzzy Logic Systems and Feed Forward Neural Networks. *IEEE Transactions on Neural Networks* 11(2):356–365.

100. Zadeh, L.A. 1996. Fuzzy Logic D Computing with Words. *IEEE Transactions on Fuzzy Systems* 4(2):103–101.

101. Ganguli, R. 2003. Application of Fuzzy Logic for Fault Isolation of Jet Engines. *ASME Journal of Engineering for Gas Turbine and Power* 125(3):617–623.

102. Kosko, B. 1997. Fuzzy Engineering. Upper Saddle River, NJ: Prentice-Hall.

103. Zadeh, L.A. 1975. The Concept of a Linguistic Variable and Its Application to Approximate Reasoning. *Information Sciences* 8(2):199–251.

104. Abe, S., and Lan, M.S. 1995. A Method for Fuzzy Rules Extraction Directly from Numerical Data and Its Application to Pattern Recognition. *IEEE Transactions on Fuzzy Systems* 3(1):18–28.

105. Wang, L.X., and Mendel, J.M. 1992. Generating Fuzzy Rules by Learning from Examples. *IEEE Transactions on Systems, Man and Cybernetics* 22(6):1414–1427.

106. Pawar, P., and Ganguli, R. 2003. Genetic Fuzzy System for Damage Detection in Beams and Helicopter Rotor Blades. *Computer Methods in Applied Mechanics and Engineering* 192(16–18):2031–2057.

107. Gao, J., and Lu, M. 2005. Fuzzy Quadratic Minimum Spanning Tree Problem. *Applied Mathematics and Computation* 164:773–788.

108. Shieh, C.S. 2002. Genetic Fuzzy Control for Time-Varying Delayed Uncertain Systems with a Robust Stability Safeguard. *Applied Mathematics and Computation* 131:39–58.

109. Verma, R., Roy, N., and Ganguli, R. 2006. Gas Turbine Diagnostics Using a Soft Computing Approach. *Applied Mathematics and Computation* 172(2):1342–1363.

110. Leonard, J., Kramer, M., and Unger, L.H. 1992. Using Radial Basis Functions to Approximate a Function and Its Error Bounds. *IEEE Transactions on Neural Networks* 3:624–627.

111. Hickernell, F.J., and Hon, Y.C. 1999. Radial Basis Function Approximations as Smoothing Splines. *Applied Mathematics and Computation* 102:1–24.

112. Li, X. 1998. On Simultaneous Approximations by Radial Basis Function Neural Networks. *Applied Mathematics and Computation* 95:75–89.

113. Goldberg, D. 1998. *Genetic Algorithms in Search, Optimization, and Machine Learning.* Reading, MA: Addison-Wesley.

114. Hajela, P. 1999. Nongradient Methods in Multidisciplinary Design Optimization—Status and Potential. *Journal of Aircraft* 36(1):255–274.

115. Wei, L., and Zhao, M. 2005. A Niche Hybrid Genetic Algorithm for Global Optimization of Continuous Multimodal Functions. *Applied Mathematics and Computation* 160:649–661.

116. Adams, D.E., and Nataraju, M. 2002. A Nonlinear Dynamics System for Structural Diagnosis and Prognosis. *International Journal of Engineering Science* 40:1919–1941.

117. Roy, N., and Ganguli, R. 2005. Helicopter Rotor Blade Frequency Evolution with Damage Growth and Signal Processing. *Journal of Sound and Vibration* 283(3–5):821–851.

118. Krupka, R.M., and Baumanis, A.M. 1965. Bending-Bending Mode of Rotating Tapered Twisted Turbo Machine Blades Including Rotary Inertia and Shear Deflection. *Journal of Engineering for Industry* 91:1017.
119. Thomas, J., and Abbas, B.A.H. 1975. Finite Element Model for Dynamic Analysis of Timoshenko Beam. *Journal of Sound and Vibration* 41:291–299.
120. Rao, S.S., and Gupta, R.S. 2001. Finite Element Vibration Analysis of Rotating Timoshenko Beams. *Journal of Sound and Vibration* 242(1):103–124.
121. Takahashi, I. 1999. Vibration and Stability of Non-Uniform Cracked Timoshenko Beam Subjected to Follower Force. *Computers and Structures* 71:585–591.
122. Hou, J., Wicks, B.J., and Antoniou, R.A. 2000. An Investigation of Fatigue Failures of Turbine Blades in a Gas Turbine Engine by Mechanical Analysis. *Engineering Failure Analysis* 9:201–211.
123. Fatemi, A., and Yang, L. 1998. Cumulative Fatigue Damage and Life Prediction Theories: A Survey of the State of the Art for Homogeneous Materials. *International Journal of Fatigue* 20(1):9–34.
124. Kujawski, D., and Ellyin, A. 1984. A Cumulative Damage Theory of Fatigue Crack Initiation and Propagation. *International Journal of Fatigue* 6(2):83–88.
125. Simha, K.R.Y. 2001. *Fracture Mechanics for Modern Engineering Design*. Hyderabad: Universities Press.
126. Jing, J.P., Sun, Y., Xia, S.B., and Feng, G.T. 2001. A Continuum Damage Mechanics Model on Low Cycle Fatigue Life Assessment of Steam Turbine Rotor. *International Journal of Pressure Vessels and Piping* 78:59–64.
127. Naeem, M., Singh, R., and Probert, D. 1999. Implication of Engine Deterioration for a High-Pressure Turbine-Blade's Low-Cycle Fatigue (lcf) Life-Consumption. *International Journal of Fatigue* 21(8):831–847.
128. Lawson, C.P., and Ivey, P.C. 2005. Turbo Machinery Blade Vibration Amplitude Measurement through Tip Timing with Capacitance Tip Clearance Probes. *Sensors and Actuators A: Physical* 118(1):14–24.
129. Carrington, I.B., Wright, J.R., Cooper, J.E., and Dimitriadis, G. 2001. A Comparison of Blade Tip Timing Data Analysis Methods. *Proceedings of the Institute of Mechanical Engineers: Part G* 215:301–312.
130. Kumar, S., Roy, N., and Ganguli, R. 2007. Monitoring Low Cycle Fatigue Damage in Turbine Blades Using Vibration Characteristics. *Mechanical Systems and Signal Processing* 21(1):480–501.
131. Lemaitre, J., and Chaboche, J.L. 1990. *Mechanics of Solid Materials*. London: Cambridge University Press.
132. Krajeinovic, D., and Lemaitre, J. 1987. *Continuum Damage Mechanics Theory and Application*. Berlin: Springer.
133. Xiaohua Yang, Z.J., Lit, N., and Wang, T. 1997. A Continuous Low Cycle Fatigue Damage Model and Its Application in Engineering Materials. *International Journal of Fatigue* 19(10):687–692.
134. Akula, V.R., and Ganguli, R. 2003. Finite Element Model Updating for Helicopter Rotor Blade Using Genetic Algorithm. *AIAA Journal* 41(3):554–556.
135. Gertler, J.J. 1998. *Fault Detection and Diagnosis in Engineering Systems*. New York: Marcel Dekker.
136. Friswell, M.I., and Jet, P. 1997. Structural Damage Assessment Using Advanced Signal Processing Procedures. Presented at EUROMECH 65 International Workshops DAMAS 97, Sheffield, UK.

137. Greaves, R.W., and White, E.R. 1987. *An Overview of Airborne Vibration Monitoring (AVM) Systems*. SAE Technical Paper Series 871731.

138. Harker, R.W., and Handelin, G.W. 1990. *Enhanced On-Line Machinery Condition Monitoring through Automated Start-up/Shutdown Vibration Data Acquisition*. American Society of Mechanical Engineers 90-GT-272.

139. Hartranft, J.J. 1995. *Description of a Vibration Diagnostic Trending Approach for a Condition Monitoring System for the LM2500 Gas Turbine*. American Society of Mechanical Engineers 95-GT-373.

140. Oberholster, A.J., and Heyns, P.S. 2004. On-Line Fan Blade Damage Detection Using Neural Networks. *Mechanical Systems and Signal Processing* 20(1):78–93.

141. Larsen, G.C., Hansen, A.M., and Kristensen, O.J.D. 2002. Identification of Damage to Wind Turbine Blades by Modal Parameter Estimation. Report Riso-R-1334 (EN). Roskilde, Denmark: Riso National Laboratory.

Index

Milton Keynes UK
Ingram Content Group UK Ltd.
UKHW031147141024
449569UK00024B/996